The Contrary Forces of Innovation

The Contrary Forces of Innovation

An Ethnography of Innovation in the Food Industry

Thomas Hoholm
BI Norwegian School of Management, Norway

© Thomas Hoholm 2011

All rights reserved. No reproduction, copy or transmission of this publication may be made without written permission.

No portion of this publication may be reproduced, copied or transmitted save with written permission or in accordance with the provisions of the Copyright, Designs and Patents Act 1988, or under the terms of any licence permitting limited copying issued by the Copyright Licensing Agency, Saffron House, 6-10 Kirby Street, London EC1N 8TS.

Any person who does any unauthorized act in relation to this publication may be liable to criminal prosecution and civil claims for damages.

The author has asserted his right to be identified as the author of this work in accordance with the Copyright, Designs and Patents Act 1988.

First published 2011 by
PALGRAVE MACMILLAN

Palgrave Macmillan in the UK is an imprint of Macmillan Publishers Limited, registered in England, company number 785998, of Houndmills, Basingstoke, Hampshire RG21 6XS.

Palgrave Macmillan in the US is a division of St Martin's Press LLC, 175 Fifth Avenue, New York, NY 10010.

Palgrave Macmillan is the global academic imprint of the above companies and has companies and representatives throughout the world.

Palgrave® and Macmillan® are registered trademarks in the United States, the United Kingdom, Europe and other countries.

ISBN: 978–0–230–28366–4

A catalogue record for this book is available from the British Library.

A catalog record for this book is available from the Library of Congress.

10 9 8 7 6 5 4 3 2 1
20 19 18 17 16 15 14 13 12 11

Transferred to Digital Printing in 2014

This need to make, to create, to invent is, no doubt,
a fundamental human impulse. But to what end?
Paul Auster

Contents

Part III

Illustrations

Figures

Acknowledgements

I am indebted to Håkan Håkansson and Per Ingvar Olsen at Norwegian School of Management BI for stimulating and challenging discussions on the topics of this study. I am also grateful to Tor Hernes (CBS), Lars Gunnar Mattsson (SSE) and Claes-Fredrik Helgesson (Linköping) for insightful comments at different stages of this project. I have through these years benefited greatly from discussions with – among others – Luis Araujo, Elizabeth Shove, Davide Nicolini, Daniel Beunza, Hans Kjellberg and Alexandra Waluszewski. Thanks to Alessia Contu for mind-opening conversations in Lancaster, to Sebastian Bringsværd – we started this journey together, to Beate Karlsen, Andreas Brekke, Anne Louise Koefoed and Gard Paulsen for many rewarding conversations, to Bjørn Erik Mørk, for great collaboration, and to all my colleagues at the Department of Innovation and Economic Organisation.

This study was done at the Center of Cooperative Studies, Norwegian School of Management, and made possible by all the people at Tine and Bremnes Seashore that opened up their organisations for me. Thank you!

Last, but not least, to my beloved wife Camilla and my children Sebastian, Rasmus, Oliver and Linus. You have been extremely patient with me. Love.

Part I

1
Understanding Innovation As Process

Innovation processes will always be characterised by tensions, controversies and conflicts. While conflict, of course, is not very pleasant for any participating actor, it is a precondition for learning, change and innovation within and across organisations. Friction is likely to produce resistance as well as release creativity. On the one hand, tensions tend to arise when status quo is challenged, or when people's positions are threatened. On the other hand, tensions can release new thinking, and produce action towards goals novel to the organisation. This is the point of departure for this study of how knowledge and technology are developed and commercialised in industrial settings.

I think we need to both acknowledge the (often constraining) historicity and 'pre-existence' of things, and understand how this also enables action in certain, and often multiple (but not unlimited), directions. In the literature on organisation and innovation processes and practices, this is known by a number of useful terms: contingency, situatedness, relationality, heterogeneity and co-creation are but a few. As opposed to the more deterministic conception of path dependency within the organisational literature, the investigated processes of my study might be viewed as efforts towards 'path creation' (Garud and Karnøe, 2001); that is, how a small product development project possibly played a part in the larger quest for industrialising a new resource. Innovation processes that span different settings (sectors, firms, networks, markets) involve complex operations of translating materials, technologies, knowledge, work practices, ideas and interests. This is not only related to the technical development of innovations, but also to the creation of markets, or users, of innovations. Hence, there are interactions between technical and conceptual development on the one hand, and the response from (potential) customers/users on the other. These tensions between the

random and the intentional, and the interaction between heterogeneous actors and resources, produce complexity and contingency to an
extent that *uncertainty* – or lack of knowledge – to some degree will
always be a challenge in innovation processes (Kline and Rosenberg,
1986; Pavitt, 2005; McMullen and Shepherd, 2006).

Innovation is a random phenomenon, it happens by chance.
Innovation can be managed, organised and strategised. Through my
journey of researching innovation, learning and organising, I have
repeatedly met both these points of view among fellow researchers.
Which is right? If innovation is random, then it has hardly any relevance
to research, as there would be little to learn from each particular case,
with regard to either successes or failures. Further, in practice, we see
how actors actively and repeatedly engage in innovative projects. Why
would they do so, if they had not had some kind of experience or belief
that they could influence the outcome? If, on the other hand, innovation can be managed, controlled so to speak, then it is rather strange
that research had not detected the laws or mechanisms of innovation
long ago. In addition, we see how many innovation projects fail in practice and, furthermore, end up being completely different from the initial intention/idea. So, if innovation is neither random, nor predictable,
what is it then? What is this space in between, where innovation processes emerge, and sometimes succeed? Is it just about identifying what
factors are manageable and which are outside our control? I think it is
more difficult than that.

Innovation in the making: a case study

This study is a contribution to unpacking this black box of innovation,
through its investigation of innovation processes and practices in one
small sector of the business world: a Norwegian agricultural cooperative, Tine, and its counterparts in the agricultural and biomarine industries. The interaction is mapped out via an ethnographic case study
of the organising of innovation processes, meaning the development
and commercialisation of hybrid technologies and products between
aquaculture and agriculture. What I am describing is the emergence
of a possibility – the possibility of industrialising fish – and several
(very early) attempts at doing so. From the scientific development of
fish farming during the last three decades, fish has now become a
controllable resource – the volumes, qualities and properties can now
be manipulated and controlled on large scale. The domestication of
salmon during the past few decades represents a breakthrough in the

large-scale production of biomarine food, enabling the control of quality and quantity of a resource previously characterised by variable access (catch). Norwegian scientific communities, together with Norwegian fish farmers, have been at the forefront of this development, and fish is often described as the second most valuable resource in Norway – and hence what Norwegians are supposed to live off after the oil era. One of the most important and interesting questions emerging from the combination of these factors, is how the fish industry will be organised commercially. Will it continue to sell fish in bulk as a 'raw material' to processors and distributors globally, in spot-price auction-like markets? Or will the domestication of fish trigger a restructuration, not only of the sizes of fish farming actors, but also of the technical and commercial practices of the industry? In other words, will there be an emergence of a food production industry based on fish, similar to what has existed for more than a hundred years within agriculture, alternatively a convergence of agri- and aquaculture?

Within this setting, I have studied a particular attempt at developing and commercialising new products from fish, a set of processes that started out with ideas of fermentation of fish (making 'salami' of fish) on the one hand, and new technologies for processing fresh salmon on the other. With a diverse set of involved actors and a relatively radical innovation to be developed, the scene was set for ample opportunities to observe the 'contrary forces' of innovation in practice. A creative scientist tried to use an agricultural technology, fermentation, to help industrialise Norway's abundant raw material resource, fish. An agricultural cooperative sought new opportunities for business in the biomarine area, based on their established knowledge and technologies of industrial production and marketing of dairy products. In addition, a fish farm looked for ways to create more economic value from its new technologies for processing salmon of supreme quality. In *principle*, these could be viewed as separate projects, not having too much to do with each other. Still, in *practice*, these projects became partly integrated and partly competing, and some of the most interesting aspects of this study are located in the interaction between these projects, and their subsequent combination into a single brand, organised in a joint venture. Salami is made by letting tribes (or 'cultures') of lactic acid bacteria process and mature the meat, and it was exactly this traditional salami recipe that served as the starting point for the project. The product that during this process came to be named 'SALMA Cured' was, in the most basic sense, a combination of fish as raw material and fermentation as technology. Therefore, it has informally come to be called a salami, or sausage, among the project participants, and 'fish salami' will also be used in this book when referring to the more general idea of this invention, rather than to its various specific appearances and names throughout the project. However, at the time this book was being written, the product that the consumer could find in an increasing number of restaurants and supermarkets was very different: 'SALMA Fresh', loins of salmon of high-end quality. Still, as this study documents, the story behind this product is a lot more complex than the neatly designed transparent package of high-end salmon would suggest. It is a socio-material drama consisting of several partly overlapping episodes, in which the actors struggled to cope with a set of challenging questions: how do you develop, realise and commercialise a food product that no one has heard of before – a product that falls entirely between established categories, both as it is perceived by consumers, and as it is organised in supermarkets and restaurants? Further, how do you make processing technologies work when they are applied to a new material?

During the initial period of farming and domesticating fish, it was not easy for single actors in the fish industry to develop practices outside their existing system of exchanging and distributing fish. This industry was raw material-oriented, and lacked competence in processing, distribution and marketing. Due to the recent consolidation of the industry, a few of the largest actors could perhaps succeed through renegotiating deals with their biggest customers, and by dramatically expanding their competence on industrial production, product development and marketing. Yet, even then, it would be very demanding, since they could not ensure this would bring about added profitability. An alternative would be to integrate agricultural and aquacultural actors in a joint effort to make this new resource available through the agricultural industrial system.

This case study demonstrates such an attempt to industrialise fish on agricultural premises. It is a case of making 'meat' of fish, an innovation project trying to impose agro-industrial practices on fish. In a way, this is a study of micro-practices: a partly ethnographic and partly historical case study of one project within a portfolio of several biomarine projects that Tine had initiated during the last decade. I closely examined these practices and processes of innovation, as they related to R&D, strategic management, production, distribution and marketing, describing the 'heterogeneous engineering' that took place across a number of professions, materials and collectives that made the project come alive in spite of trials, setbacks and shifts; processes that radically altered the project and its initial object(-ive). On the other hand, the ongoing micro-practices of realising this one concept should be viewed as a mutually constitutive part of what might become new industrial practices between agri- and aquaculture. It is an example of how a changing 'macro-structure' creates new opportunities, but also a bottom-up perspective of how new industrial and market practices are made. Scientists, technologists and suppliers from both sectors were enrolled into the project, bringing in their own knowledge and perspectives, raw materials from both domains were recombined in innovative and technically advanced ways, and the application of agricultural marketing practices to the products of aquacultural materials was attempted.

The case study is then used to analyse and discuss the organising of industrial innovation processes in practice. Innovation is viewed in this book as something that happens between sectors, knowledge areas, organisations and networks. This 'in-between-ness' not only opens up for new combinations, and hence, new business opportunities, but it also creates fundamental organising and marketing problems of ambiguity, destabilisation and complexity. In this respect, this case study is not unique, as arguably all innovation happens across many boundaries, whether they be between networks, sectors or organisations. Thus, this book presents a detailed and situated study of a general phenomenon, namely industrial innovation. Innovation in this setting implies the development and commercialisation of knowledge and technology that then manifest themselves in new products and ingredients, and in new practices that cross, or reorganise, the traditional boundaries between agri- and aquaculture. A dual dynamic came to the fore in my analysis of the case, between mobilising actor-networks on the one hand and exploring knowledge on the other. These 'sub-processes' of the larger innovation process sometimes drew on each other, at other times they did not interact at all and sometimes they came to confront each other – with potentially serious implications for the future of the innovation. In growing out of the empirical study, while also receiving clear inspiration from my methodological and theoretical frameworks, this dual dynamic became a 'conceptual model' from which I have structured a theorising discussion on innovation processes.

Situating the study

'Innovation' is a popular, yet problematic word within different public discourses today; politicians compete in proclaiming their nation or region to be becoming more innovative than their neighbours; business associations and companies complain about the lack of suitable conditions for and efforts towards innovation in their environments; and economists, sociologists and technologists all claim to know something about what constitutes the keys to innovation. It is a rapidly growing field of research, particularly on aggregated levels, such as systems of innovation (Edquist, 2001; Lundvall et al., 2002), clusters (Porter, 1998) and various network-theories (e.g., Powell et al., 1996; Håkansson and Waluszewski, 2007a). Nevertheless, we still do not know enough about the 'content' of *innovation processes*, and we are unable to account properly for how innovation evolves in practice (Van de Ven et al., 1999; Garud and Karnøe, 2001; Gupta et al., 2007).

When studying innovation processes, one of the most basic (and, ultimately, philosophical) questions is whether innovation is best understood in terms of (1) *changing* something stable, or (2) *stabilising* something fluid. In the first case the world is viewed as rather stable, hence the problem (or the exception) is how to change. Relationships are analysed as structures, and resistance to innovation and change is often explained as 'inertia'. Systems of innovation (Edquist, 2001), Rogers' theory of diffusion (1995) and organisational neo-institutionalism (Scott, 1995), are examples of perspectives we could put under this category. This point of departure has been shared within mainstream organisation and innovation research during the past few decades of circling the problems of change and how to loosen the grip of inert and solid 'structures', hence triggering a call for process-based research (Tsoukas and Chia, 2002; Van de Ven and Poole, 2005; Hernes, 2007). Thus, if our interests are located within the problem of how things come about, the complex emergence and realisation of novelty in socio-material interaction, the second line of reasoning is suggested as being a better-suited approach. It seems to me, for instance, that our understanding of the emergence and stabilisation of inter-organisational relationships and networks is still quite limited. Network *change* is not necessarily the problem, *stabilisation* is; that is, how learning or innovation results in new common and stable practices (or not). In this study, I examine *innovation processes* in which relationships are initiated, changed and broken. In a large segment of the research literature on (inter-) organisational relationships, networks and systems, these factors are analysed as if they are given and stable entities. This represents a problem if our research interests include understanding how relationships and networks come about in the first place, and how they evolve over time. In the second line of reasoning, the world is viewed as constantly changing, the problem (and the exception) being how to stabilise anything.[1]

Relationships are analysed as recursive processes of interaction (Law, 1994), and the concept of 'friction' is suggested to explain resistance to – and sources for – innovation (Håkansson and Waluszewski, 2001a,b). Actor-network theory (ANT) and other science and technology studies (STS) approaches, the industrial network approach (IMP[2]) and a few process approaches to organisation and innovation, such as Weick (1995), Van de Ven et al. (1999) and Tsoukas and Chia (2002), are representative of this as a basic assumption.

In particular, I am positioning the study within a set of emerging fields of research aiming towards understanding innovation and markets from process and practice points of view. While science and technology studies have long investigated knowledge production and technology development, we have recently witnessed a growing interest in following technology and knowledge further into the economic and industrial world. On the business side, Van de Ven and others have contributed to developing a process view of *innovation management and organisation*, while the IMP group has done studies of innovation in *industrial networks*. While these fields of research are partly overlapping, they have also left some 'blank spots' in relation to industrial innovation. In relation to STS, the study shows how the pragmatics of business (which may be understood as a set of practices connected to economic theory, consumer practices, industrial networks, marketing/branding practices, etc.) may compromise the technological passions and interests driving technical innovation, making science less 'pure', less hegemonic and less seductive. In the encounter between inventions of technoscience and the tough 'realities' of business, what remains of the initial innovation is an open question, along with how it is combined with existing business practices in order to find use(-rs), and how it takes part in reconfiguring and reconstituting those practices. In relation to innovation management and industrial networks, the study shows how technoscience, and its creative chaos (of ideas and direction) and rigidity (of method), may serve as precondition and resource for creating new commercial practices.

Below, I briefly introduce these theoretical perspectives, to situate my work within and between them, to account for my interpretative basis when analysing the case and to prepare the reader for the theorising discussion thereafter. However, I have chosen to present specific contributions that I both draw on and challenge in Chapter 6, which is close to the theorising discussion. Moreover, the presentation of theoretical contributions below is not symmetrical. A basic assumption in this book is that research strategies from the particular field of science and technology studies called actor-network theory – or 'sociology of translation' – are also particularly useful for furthering our understanding of industrial innovation. Therefore, I have allocated more room for presenting this perspective.

Innovation management and process

Kline and Rosenberg (1986) did groundbreaking work in pinpointing and researching the intertwining of technology and economy in innovation processes. They claimed that economists had black-boxed the process of technical transformation, while technologists often failed to take the 'external forces of the marketplace' into consideration. Innovation, from this perspective, is a complex and uncertain process, and an 'exercise in the management and reduction of uncertainty' (ibid.: 276). However, their distinction between the (technical) innovation process and its interaction with an external market environment did not go far enough, despite introducing feedback-links as central elements in their 'chain-linked model'. Science and technology were ascribed interactivity, but Von Hippel (1988) a few years later became the major proponent for considering the market – in the shape of users/ lead-users – as internal to the process, granting full interactivity also to the 'users', and influencing the innovation and its fate by using, modifying and/or rejecting it. The inclusion of heterogeneous actors and resources in this way has certainly produced a complex and contingent view of innovation processes. This is underscored by Pavitt:

> Innovation processes differ in many respects according to the economic sector, field of knowledge, type of innovation, historical period and country concerned. They also vary with the size of the firm, its corporate strategy or strategies, and its prior experience with innovation. In other words, innovation processes are 'contingent'. (Pavitt, 2005: 87)

Contingency, in this sense, implies that innovation processes always carry a degree of uncertainty, and that uncertainty is reduced (although rarely removed completely) by learning from and building on experience and continuous feedback from the market. According to Pavitt, only two aspects of the innovation process are generic, 'coordinating and integrating specialised knowledge, and learning under conditions of uncertainty' (ibid.: 109). In reviewing the role of uncertainty in entrepreneurship research, McMullen and Shepherd identified three different kinds of uncertainty, all related to a combination of time and novelty: 'state uncertainty' denotes an unpredictable environment, 'effect uncertainty' denotes the uncertain effect of the (future state of the) environment to the organisation and, finally, 'response uncertainty' denotes lack of knowledge of – and unpredictability of the effect of – response options (McMullen and Shepherd, 2006: 135). While being argued from a dualist point of view, of *internal* uncertainty in the face of *external* change, these kinds of uncertainty are also relevant to an *interactive* view of innovation, although with the effect that the conceived uncertainty would probably expand radically.

McMullen and Shepherd (2006) also discuss whether entrepreneurial 'opportunity' should be viewed as something being (objectively) *discovered*, or as something *enacted* and co-created by the actors involved in interpreting knowledge/information in the situation, thereby resembling the distinction between realist and constructivist ontologies. Both emphasise that opportunities are characterised by uncertainty and that this is central to understanding entrepreneurial action. Yet, where the realists' uncertainty is mainly about availability (and sometimes interpretation of) objective knowledge, constructivists' uncertainty relates to the enactment of perceived opportunity by the involved actors, hence multiplying the sources of uncertainty. According to the latter view, the 'out there/in here' distinction between actors and environments dissolves. Uncertainty, then, comes from the unpredictability of interaction, both between human actors (individuals or constellations), and between humans and material elements ('natural', technical and textual). This is in line with Kline and Rosenberg's (1986: 276) emphasis on our need to improve our understanding of the 'management and reduction of uncertainty'. The degree of uncertainty in innovation processes are, according to Kline and Rosenberg (1986: 294), 'strongly correlated with the amount of advance' and, according to Pavitt (2005: 105), the risk of failure in innovation processes will increase 'with the number of practices and competencies that need to be changed'. Such 'radical' innovation processes are equated with processes of learning, where 'overplanning' may distort the process because the future cannot, by definition, be fully known. Furthermore, the 'false summit effect' – or repeatedly finding new mountaintops behind the one that was believed to be the real summit – produces a kind of uncertainty that cannot be planned for in detail. However, innovators tend 'to underestimate the number of tasks that must be solved and hence also the time and costs' (Kline and Rosenberg, 1986: 298). In addition to the technical side, then, uncertainty is increased by rising development costs, resistance to radical innovation, financial risks and coupling of the technical and the economic. Hence, economic forces, technical knowledge and consumer demand need to be closely connected during innovation processes.

From this, we see that innovation consists of contingent processes, stemming from interaction between science, technology and markets, thereby producing high levels of uncertainty. Hence the 'management of uncertainty' is one of the crucial tasks for participants in such processes. Pavitt (2005), then, found that coordination of knowledge and learning were the two generic aspects of innovation processes. It also seems clear that the more change that is needed, the greater the uncertainty will be. However, he notes that more research is needed for improving our understanding of these dynamics.

The longitudinal comparative study of innovation in the MIRP-study (Minnesota Innovation Research Project), as reported in Van de Ven et al. (1999), has become an obligatory point of passage for anyone studying innovation processes. Their systematic gathering of longitudinal data from a number of cases and sophisticated analysis of complex patterns has left a lasting mark on the study of innovation. Their main thesis is that the common pattern of all innovation processes 'is a nonlinear cycle of divergent and convergent behaviours that may repeat itself over time and reflect itself at different organisational levels' (Van de Ven et al., 1999: 213). They found this to be the case independent of the big diversity of paths and outcomes of the processes studied. Linear stage models as well as random models are disputed; instead, they argue for innovation as 'emergent process' based on nonlinear dynamics, in which sensitivity to initial conditions and the ability to manage complexity are viewed as being crucial for success. A number of important implications are drawn out of the study. Again, learning is viewed as being a central aspect of the process, where 'learning by discovery' is understood as 'an expanding and diverging process', and learning by testing as 'a narrowing and converging process' (ibid.: 203). In turn, these ways of learning are explained as being dependent on each other in a continuous cycle. While confirming the uncertainty of outcomes and thus acknowledging that innovation processes cannot be controlled, Van de Van et al. still argue for *managing* innovation (in terms of navigating rather than controlling). In their study, they found that managers' performance criteria shifted over time, in relation to outcome, process and input, and in line with the changing needs of the innovation process and the unexpected events that occurred. Such changes 'triggered innovation managers and entrepreneurs to search and redefine their innovation ideas and strategies' (ibid.: 42). Beunza and Stark (2004) and Howard-Grenville and Carlile (2006) confirm this argument, although from a more political point of view, showing how the negotiation of evaluation criteria is fundamentally a political process through which power relations are (re-)constituted. This activates a need for the 'management of paradox', in which highly effective organisations are able to perform 'in contradictory ways to satisfy contradictory expectations' and 'ambiguity in goals' (Van de Ven et al., 1999: 12). Thus, in order to succeed with innovation, there is a need to acquire power via coalition building. This is accomplished through selling the project to various stakeholders in different ways during the process. Ambiguous and uncertain situations, like early phases of innovation processes, 'require a pluralistic power structure of leadership' (ibid.: 124). This increases the chances for technological foresight, while also decreasing the chance of oversight. Including a diversity of views and conflicts is viewed as constructive for divergent innovation processes, and serving as 'checks and balances with each other'. On the other hand, unitary, single-vision and hierarchical leadership tends to restrict creativity and (necessary) deviant behaviour. Coalitions, or networks, tend to grow over time, resulting in a complex network 'engaging in a series of transactions necessary to move the innovation forwards' (ibid.: 50). This networking process is non-linear, characterised by 'numerous bargaining, commitment, and execution events' of the inter-organisational relationships. After some time, the networking may reach a point of 'self-organising criticality', 'wherein the relevant unit of analysis becomes the web and not the dyad'. Moreover, interactions in the web were often more influenced 'by activities occurring in other dyads than by the internal logic of the dyad itself', making it necessary to include web-level effects in the analysis of innovation processes (ibid.: 148). Lastly, Van de Ven et al. found that 'Innovation uncertainty decreases over time as system functions that define key technical and institutional parameters for the innovation emerge' (Van de Ven et al., 1999: 172). However, in my case study, the reduction of uncertainty was less about getting system functions, or institutional arrangements, in place, than about a process of radical simplification – of stripping down the innovation itself in order to get adaptability – and thus momentum – towards relations and patterns of distribution and use (see Chapter 6). In the continuation of the MIRP-study, many questions on innovation processes still remain to be answered. What are the dynamics driving the divergence, convergence and interaction between the two? What is manageable and how? And, in the face of the complexity of networked interaction, how is it that such processes (sometimes) stabilise? How do 'system functions' and 'institutional arrangements' come into place? This book might be read as a contribution towards understanding the content of innovation processes.

In a continuation of the MIRP-study, Garud and colleagues have contributed with new insights into certain aspects of innovation processes. Drawing on a 'socio-cognitive' model of technology evolution, Garud and Rappa focus 'on the relationship between the beliefs researchers hold, ... the artifacts they create, and the routines they use for evaluating how well their artifacts meet with their prior expectations' (1994: 344). A distinction is made between two different cyclical processes: one in which 'evaluation routines designed to judge specific artifacts begin reinforcing researchers' beliefs', and another – institutionalisation – which develops a 'common set of evaluation routines that can be applied to all technological paths'. Still, these processes represent a shared reality. Garud and Rappa observed how beliefs were externalised by creating routines, which in turn were used to evaluate the technology in a self-reinforcing circle. However, the influence went both ways, as the technical artefacts also had severe impact on what kinds of evaluation routines could be employed. The lack of stability in the relationship between the emerging innovation and evaluation routines/criteria thus makes new technology particularly precarious during its early stages of development and use. In their discussion of technology assessment, Garud and Ahlstrom further Garud and Rappa's argument, in that the criteria of evaluation are explained as being negotiated between involved technologists. Sets of evaluation criteria form 'frames of reference',[3] which create a self-fulfilling prophecy by rendering the researchers blind to alternative technological trajectories (1997: 27). This also creates a need for 'outsider' evaluations to broaden the range of discussion and challenge taken-for-granted criteria. On the 'positive' side of technologists' blinkers, Garud and Karnøe (2001) have investigated the role of (and space for) agency in shaping new industrial practice, or, in other words, shaping new technical paths. Combining thinking both from innovation management and STS literature, they argue that 'mindful deviation' is a central characteristic of how entrepreneurs contribute to 'path creation' and thus towards implementing new ideas in the economy. Van de Ven et al.'s (1999) model of innovation as 'punctuated learning', and Garud and Karnøe's concept of path creation (2001) will be discussed in more detail in Chapter 6.

The industrial network approach

Throughout the 1970s, a network-oriented perspective on markets and marketing, especially related to industrial markets, emerged out of a set of Swedish research projects (Mattsson and Johanson, 2006). In developing a relational/interactive perspective, they challenged existing views on market practices. Suppliers and customers were depicted as mutually dependent and embedded in long-term relations, which then also served to depict markets as consisting of interconnected relations. Firm strategy was understood to develop in network interaction, which thereby blurred the boundaries of the firm, and was viewed as more as an effect of learning than of planning. Moreover, product development and innovation were often integrated into these relationships (ibid.: 261–5). The proponents of this perspective are often referred to as the IMP-group,[4] the 'markets-as-networks approach', the 'interaction approach' or the 'industrial networks approach', and I will use the abbreviation 'IMP' below for the sake of simplicity. The IMP perspective involves numerous rich case studies of both relations and dynamics in networks, and has challenged the traditional view of business markets, arguing that management of relationships in inter-organisational networks, rather than internal allocation of resources, is the crucial strategic issue (Håkansson and Snehota, 1989).

So, how is interaction conceived of within this 'interaction approach'? What is the driving force, the rationale behind actors' choices to interact? First, sometimes actors do not have a choice; due to their social and material embeddedness in relationships, interaction is necessary to survive or to get anything done at all. Still, when one has a choice, interaction might be sought out for different reasons, including problem solving, learning, innovation, efficiency or cost reductions (Ritter and Ford, 2004). Ford et al. (2003: 7–8) have employed 'networking' as a synonym to interaction in business networks, claiming that all companies are networking, which means 'suggesting, requesting, requiring, performing and adapting activities, simultaneously'. Individual and isolated action is therefore irrelevant, as networking involves 'reaction to the actions of others and all of it will have to take into account the reactions of others'. In other words, interaction is characterised by mutual processes that are never one-sided; rather, interaction is always a collective achievement that affects all interacting parties.

Within the framework of IMP, interaction is explicitly regarded as 'the essential analytical concept at the heart of a relationship and network perspective of business markets' (Medlin, 2002: 1). As mentioned previously, this view emerged from an intense critique of traditional economic paradigms and their myths of individual action, independence and completeness. Instead, interaction is understood as being an interdependent process, in which 'no company *alone* has the resources, skills or technologies that are necessary to satisfy the requirements or solve the problems of any other' (Ford et al., 2003: 2). Moreover, each participant involved in interaction takes different 'pictures' of the network with him/her. The main research issues within IMP are summarised by Ford et al. (ibid.: 8–11) in three 'network paradoxes':

(1) A company's relationships are the basis of its current operations and development. But those relationships also restrict that development.
(2) It is equally valid to say that a company defines its relationships or that a company is defined by those relationships.
(3) Companies try to control the network and want the benefits of control, but control has its problems and when it becomes total, it is destructive.

This means that networks consist of relationships which then consist of interactions. The focus is often on the restrictions of interacting in networks; that is, actors within industrial networks cannot act as they want, since they are embedded into complex webs of relationships (Håkansson and Ford, 2002: 135). Networks are often considered to be quite stable and difficult to change. The reason for this 'heaviness' is explained as being a result 'of complex interactions, adaptations and investments within and between the companies over time' (ibid.: 133). However, since changes in one relationship are likely to affect all connected relationships, these relationships might also be the source of considerable change in the network, thereby making understanding networks crucial for understanding interaction:

> Thus, no one interaction, whether it is a sale, purchase, advice, delivery or payment can be understood without reference to the relationship of which it is a part. Similarly, no one relationship can be understood without reference to the wider network. (Ibid.: 134)

Such interaction in industrial relations and networks may clearly be called a 'process perspective' on business practice, and a process perspective is dependent on including time in its analysis. Johanson and Mattsson (1987) elaborated interaction into two categories, namely exchange and adaptation. In order to capture the dynamic (or interactional) side of these categories, time needs to be included as a factor (Medlin, 2002; Dubois and Araujo, 2004; Ritter and Ford, 2004), because there seems to be a difference where *exchange* happens in the present, while *adaptations* are being 'planned in the present, but exist as changes to resource ties and activity links in the future' (Medlin, 2002: 7), and both concepts are closely related to both past experience, present interaction context and future expectations. Altogether, this resembles a research perspective investigating 'the social creation of reality through interaction' over time (ibid.: 4). Due to the uncertainty and indeterminacy of the future, interacting firms need to arrive at some kind of shared intention in order to achieve their goals. On the other hand, a full sharing of future scenarios would be a utopian ideal, as each actor will bring different views and experiences into the relationship. From this, we see that even on a 'purely' social level, network interaction is heterogeneous, where 'different perspectives of reality interact in a "relationship dynamic"' (ibid.: 9). However, heterogeneity of interaction becomes even more important if we include non-human elements into our analysis. Human and organisational actors do not interact in 'empty space'; on the contrary, they interact both with and through numerous artefacts, such as technologies, texts and other material resources. In the IMP literature, this aspect has been studied under the label of 'resource interaction'.

Some authors have put resources (Waluszewski, 2004; Håkansson and Waluszewski, 2001a) and, more specifically, knowledge, (Araujo, 2003; Håkansson and Waluszewski, 2007a) at the centre of their analyses of inter-organisational interaction. In such interaction between resources, possibilities for new solutions are created, and old resource combinations are confronted with new alternatives, which produces variation (Waluszewski, 2004: 146). Hence, if action is explained in relational terms – as interaction – agency is no longer necessary for action. It is no longer given *a priori* who or what will act in each specific setting under study. In their enquiry into the question of why and how technological systems, consisting of resource combinations, often hold a certain stability that makes them difficult to change, Håkansson and Waluszewski (2001b) oppose the static notion of 'inertia', instead launching the concept of 'friction'. They observed how resources often seemed to be 'cemented' upon each other, and therefore hard to change or replace, and yet resource combinations with a seemingly unlimited stability sometimes suddenly loosened. According to Araujo, 'friction captures the notion that movement of knowledge involves both transfer and transformation when different types of knowledge interact' (Araujo, 2003: 20). Friction is a relational concept, describing a force directed towards two interacting bodies. It is time-dependent, having different effects at different times. And it is transformational, in that friction not only leads to movement, but also to some kind of transformation of the interacting bodies, whether it be a change in shape, deformation or some other transformation. Friction also connects both historical and contemporary processes in being a reaction to one or more of them, and is thereby neither random nor deterministic. Based on these aspects, friction is viewed as being an 'active force' in resource interaction, intervening in any attempt to change the current embedded resource combinations, with a strong tendency to favour existing values (Håkansson and Waluszewski, 2001b: 2). In this picture, human actors are described as both advocates of change and protectors of established practice (ibid.: 3). This is a challenging task for people in innovation processes, of coping with the new without destroying the current, as change often has a number of unintended (and unpredictable) consequences in related interfaces. Yet this is also a clear limitation of human agency, as the use and value of any resource is determined by the relationships and interaction processes in which it is embedded (ibid.: 4). Innovation is thus described as 'walking in a rugged landscape', uncertain and uncontrollable, and always exposed to numerous different forces.

With regard to the distribution of effects, it is argued that forces directed towards one resource will probably also affect all of the other resources with which the focal resource interacts. Hence, effects are never merely local; they get distributed through friction with other interfaces with other resources, transforming them too. One reason for *stabilisation* effects, they argue, is that friction connects the present with the past, thereby defending earlier results and solutions. Processes of innovation, or recombination, entail examining the resources to be combined, and then trial-and-error investigations of the features that can be activated in the new interface. In other words, stability is an effect of embedding resources. This is a process of bringing histories together, with uncertain outcomes and no ideal solutions, along with the challenge of integrating the new interface with related existing interfaces (ibid.: 15). On the other hand, friction also produces *destabilisation* effects. Through friction, simultaneous processes are connected, allowing the same interface to be activated in several change processes. In this way, friction can also sometimes strengthen change (ibid.: 17). The more the focal resources are embedded within other interfaces, the more friction is produced, thereby affecting more resources, and requiring more power to initiate change (ibid.: 18).

What influences the degree of friction, then? Håkansson and Waluszewski (2001b) found 'economic heaviness' to be a conservative force, including investments in material and immaterial resources, and the combinations of these into complex webs of relationships. Resources are activated in interfaces, giving them their substance and heaviness. In this way, power has a double face – increased heaviness leads to less freedom, but also to more influence. They further argue that attempts to change are compelled to become 'economical', from the established structure's point of view (ibid.: 23). Another aspect influencing the degree of friction is *variety*, that is, a combined effect of the characteristics of the resource and its interfaces with other resources (ibid.: 20). This variety can sometimes be a problem (e.g., in large-scale industrial settings), and at other times it reveals a large economic potential.

In the above review of the innovation management literature, knowledge (or the lack thereof) was posited as the most central challenge to innovation processes. In addressing knowledge development, friction forces knowledge transfer to become more about the creation of knowledge, which not only pertains to coordinating pieces of knowledge, but also moulding them together into new forms. Furthermore, friction often leads to the uncovering of 'weak links', where the development of new knowledge is fairly easy, thus both opening up new opportunities for knowledge creation, and for the realisation that this development often will not take place at the point of initial change (ibid.: 25). It could also be noted that the 'traditional' view of knowledge creation and organisational learning has been challenged by Håkansson and Johanson (2001), who argue that learning in business networks is inclined towards differentiation, and thus complementarity, rather than the traditional view of learning as homogenisation (ibid.).

From an interaction perspective, we can outline some implications for the study of innovation processes within (and across) industrial networks. First, innovation is a highly interactive and collective issue, involving numerous actors, and 'seldom the result of one "designer"' (Håkansson and Ford, 2002: 135). Hence, an innovation is never given at the outset; it is the object of controversial and collaborative interaction often over long periods of time. Second, it is hard to innovate. Within business networks, numerous combinations are possible, enabling a potential innovation, but only as long as it is 'designed appropriately and seen to be positive by those who support its needs' (ibid.). Relationships are viewed as imposing 'severe limitations' on a single company, due to the costs of making changes, as well as all the possible unintended/indirect effects on other relationships. Third, new relationships with new counterparts are viewed as being difficult because of the existing structure that has to be taken into account. This is called the 'heaviness' of networks, and leads Håkansson and Ford (2002) and

Waluszewski (2004) to argue that innovation has a better basis if current investments, technologies and resources are included in the development, and combined with the new solution: 'Interaction demands that the "interactors" relate to the existing structure. The more the new solution can be embedded into the existing structure, the larger the economic advantage for both the supply and user sides' (Waluszewski, 2004: 147). At the same time, it is important that a place for the innovation is created, which often demands the 'interactors' to break with parts of the existing structure (ibid.). Thus, there is a question of 'matching' the new with aspects of the old, although probably it cannot be matched with all of them.

While most research on business relationships starts out with existing 'structures' of relationships, less attention has been paid to the emergence of business networks. Although it may be argued that interaction always starts out with something, it still seems appropriate to focus more on the emergent aspects, particularly in relation to innovation. Törmänen and Möller (2003: 1) claimed, at the time of writing, that we have had 'no more than a rather scant understanding of the dynamics of the emergence of business networks, and especially about the role of individual actors in this process'. In placing greater emphasis on *actors* than what has been common within the industrial networks tradition, and in aiming to study *intentional* construction of business networks, Törmanen and Möller develop the concept of 'network capabilities' to understand how networks emerge. The construction of new business nets is argued to depend on 'the ability to combine several technologies and coordinate the resources and capabilities of various actors coming from different fields', and that 'both knowledge and learning can be expected to play core roles' (ibid.: 4). They show how these network-building processes are characterised by uncertainty, calling for a conscious trial-and-error approach so as to ensure learning and adjustments along the way. They also mention another aspect of the process, namely how the management team may work to sell their agenda within the organisation, through making and keeping promises to several parties, and in this way securing access to the resources needed (ibid.: 12). Moreover, Awaleh (2008) has shown how actors may engage in 'purposeful networking'; for example, how 'networking ambassadors' in the interacting organisations may work strategically to challenge, change and align 'network pictures' in industrial relationships.

Waluszewski (2004), Håkansson and Waluszewski (2001a), Leek et al. (2003), Medlin (2002) and Dubois and Araujo (2004) all call for developing theoretical 'tools' to better cope with the dynamic aspects of networks. Mattsson (2003: 16) suggests that actor-network theory could enrich and complement the industrial networks approach by explicating how human and non-human actors are related, how social phenomena are 'performed' in emerging and heterogeneous networks, and offering a more precise methodology for studying dynamics. He also suggests that ANT could benefit from the accumulated knowledge of the IMP approach when turning their eyes towards economic and market phenomena. In particular, the IMP approach has, in my study, promoted the inclusion of established relationships in the analysis, taking more of an outside-in perspective, complementing the inside-out perspective that is more common in the field of innovation studies. Håkansson and Snehota (1995: 3) illustrate this difference in perspectives and emphasises that from a network perspective explanation of relationship development needs to 'include factors "external" to the relationship itself'. Thus, the emergence of new relationships cannot be viewed in isolation from the rest of the 'network structure'. Within the actor-network literature (see below), although having a relational basis, there has been a tendency to focus more on the relationships that are sought established, while less attention has been paid to the established set of relationships into which the innovation will be embedded. This complementation (or expansion) is especially visible in the discussion on user-interaction and mutual translation in Chapter 6, although analysing the case mostly from one of the parties' points of view. I will now account for actor-network theory and the study of emergence.

Actor-network theory and the study of emergence

The particular strand of STS called actor-network theory has sought to describe and understand the rise, continuity and fall of social networks, in viewing them as relational and heterogeneous. Such networks include not only humans, but also artefacts like technology, texts, symbols and things as actors that can mediate knowledge and participate in the social. According to ANT, the social is unstable and unpredictable, as any actor can (and often will) resist the exercise of power by others. The actors that are able to enrol other actors in their network by selling their discourse and making the other actors dependent upon their knowledge and discourse/mode of ordering will succeed in building their network, at least for a while. Actor-network theory is not so much a theory as it is an empirical and analytical methodology. As emphasised by Latour (1999a) and Law (2004), it is a way of tracing the 'world building activities', making up the social and material relations that surround us in a way that unravels what we normally take for granted.

It is fundamentally a relational and process perspective, viewing the world in constant flux and hence putting stability – and stabilisation – under scrutiny. This makes this perspective a good starting point for studying innovation, enabling the study of emergence and the problem of developing and stabilising novelty.

During the late 1970s and early 1980s, a number of ethnographic studies of scientific laboratories were carried out by researchers from sociology of science (e.g., Latour and Woolgar, 1979/1986; Knorr Cetina, 1981; Lynch, 1985). Subsequently, studies of technology development were also included (e.g., Bijker et al., 1987; Bijker and Law, 1992; Latour, 1996). Within this stream of emerging perspectives on science and technology as 'constructed', relational and shaped in and by social practice, what came to be called 'actor-network theory', or 'sociology of translation', emerged with researchers such as Bruno Latour, Michel Callon and John Law as its major proponents.[5] It was influenced by a number of other traditions, such as post-structuralism, feminism, ethnomethodology and phenomenology (Calas and Smircich, 1999: 663). Foucault's thinking has been influential, especially the 'notion of power/knowledge as power relations are produced through "actants" who perform the available discourses and practices' (ibid.). In applying the semiotic principles of relationality to all kinds of materials, ANT relates that 'entities take their form and acquire their attributes as a result of their relations with other entities' (Law, 1999: 2), thus taking a non-essentialist standpoint. Law further emphasises the uncertainty and reversibility of entities resulting from this 'material relationality'. The dualisms of agency and structure in social theory are challenged by Latour (1999a), who argues that the social is not constituted by agents and structure at all; rather, it is a 'circulating entity'. This means that reality is constituted by the circulations of transformations, being real, social and narrative at the same time, thus arguing that ANT entirely bypasses the question of 'social construction' and the 'realist/relativist' debate. To sum up ANT's ontology, it views reality as relational and multiple. Different roles and identities are constructed within different sets of relations, which demands different strategies for interaction and activity. As a consequence of this relationality, reality is also multiple. Actors must renegotiate positions and roles, mediate between expectations from different networks, and relate to truths in one network that is irrelevant in another. For the researcher, it can be problematic to define the 'object' of enquiry, because to define it would be to lose the multiple character of networks. Still, to *not* define it is to risk vagueness and loss of focus. Law describes this problem during his organisational research in a big laboratory, 'What is the laboratory? Figures? Results? A site? A lot of people? A set of plans? To define it would be to break the principle of symmetry. It is many places and many orderings, a network' (Law, 1994: 40).

The 'answer' is outlined as a struggle of representation and pragmatism, to try representing through multiple voices according to the intention of the study, recognising that telling the whole story is impossible. Thus, it is also a political question: which story do I *want* to tell? And the political aspect becomes even stronger as representation is understood as participating in the making of reality. Law (1994) make use of the Goffmanian term of 'performativity' to explain how reality is both real and produced at the same time. Performance does not mean that something is not real, but that reality is performed in socio-material relations, and therefore that performance makes reality. It is recursive. And it is not only a few 'powerful' actors that are part of these performances, it is rather the case that society 'is performed through everyone's effort to define it' (Latour, 1988: 273), scientists included. Power, identity and performance are about practice, as there are no ultimate principles determining the social, no first or last instance – the social is performed in local specific practices, in networks of networks.

Latour has, in several of his studies of knowledge production and innovation (e.g., 1988; 1996), shown that transfer or diffusion of knowledge is never just that, knowledge is never just 'flowing' or 'diffusing' through the system. Instead, he argues that the object (e.g., knowledge or an innovation) is always changing on its way. Further, he argues, it is not moving ('flowing') by itself; it is always up to the individual actor to decide whether s/he should pass it on or not, in what way and in what form. From a slightly different perspective, than Håkansson and Waluszewski's (2001b) concept of friction, Latour's (1988; 1999a) concept of *networks* in his 'model of translation' is one that emphasises 'work' more than 'net'. He argues that networks should be understood as processes of translation, association, deformation and transformation. This is because:

> the spread in time and space of anything – claims, orders, artefacts, goods – is in the hands of people; each of these people may act in many different ways, letting the token drop, or modifying it, or deflecting it, or betraying it, or adding to it, or appropriating it. (Latour, 1988: 267)

This implies that networks are *less predictable* than might be believed, and the possibility of controlling networks becomes highly questionable, because power is treated not as a cause, but as 'the consequence of an intense activity of enrolling, convincing and enlisting' (ibid.). Law (1992) adds that network ordering is also a matter of the uncertain process of *overcoming resistance*. Hence, Latour suggests the term 'translation' as more appropriate for explaining processes of knowledge transfer and innovation. Translation, then, is defined by Law as 'the process or the work of making two things that are not the same, equivalent' (1999: 8); for example, texts are often constructed as combinations of other texts, taking on the role of representing the other facts, figures, numbers, definitions, descriptions, etc. Thus, in reality, the ability to gather, simplify and represent an increasing number of materials in one is what makes knowledge mobile and influential. This is 'translation' – that is, to speak *for* and represent someone else, and to simplify and delete complex, effortful and often controversial processes. Sometimes this is called a 'black box', where only the input and the output are visible, which thereby stabilises its network. The actor-network can be said to be (more or less) stable when (and partly because) it is taken for granted, no longer questioned and problematised. ANT speaks of translation of knowledge and technology as stories of alignment and of hard work in order to stabilise the social, and has developed analytical concepts and tools for structuring these stories. The logic of actor-networking is one of negotiation, association and gathering bits and pieces into a unity.

In his work towards fleshing out pathways to a process perspective in organisation studies, Hernes (2007) sums up some of the contributions from Latour's works that are considered useful for studying innovation processes. First, no social order can endure over time except via socio-material relations (ibid.: 72). Second, these heterogeneous networks are kept together in and via recursive patterns that are repeated in time and space. Third, this means that entities (actors, resources, innovations, etc.) are the outcomes of their relations, which, by the way, never become totally stable, as they are always in the process of becoming. While Hernes (2007) portrays actor-network theory, and process studies, as perspectives that emphasise 'choice', 'play', 'experimentation', which allow 'for choices to be made while remaining open to the possibility that the outcomes will not be as expected' (ibid.: 76), he does not put as much emphasis on how actors get themselves in the position to do the experimentation and have choices in the 'first place'. In a negative sense, this means that little is said about the resistances and limitations that actors experience when trying to order things into (new) patterns, whether such resistances come from materials and (interacting) practices (Mørk et al., 2006), politics of expertise (Mørk et al., 2010) or interaction in wider networks (Hoholm and Mørk, 2009). Within structural perspectives, this immediately comes to the fore, emphasising reified 'barriers' and 'structures', and the inertia of human behaviour. Still, from a process-based and relational perspective, the explanation is inevitably different. Below, I mention two complementary explanations.

The first explanation goes to the core of actor-network theory and its materialist orientation. Law (1994) explains relative stability in terms of heterogeneous ordering. The hardest work of the actor-network, he claims, is ordering through time and space. To establish and retain performative patterns across distance, 'Some materials last better than others. And some travel better than others' (Law, 1994: 102). Durability is quite obvious; brick walls and roads last longer than speech and text on paper. Mobility is perhaps less obvious. Ordering through space is one of the most important projects throughout history. How can managers secure loyalty to and maintenance of their network in distant places? Surely, they can never fully achieve this, but through embedding ordering strategies in mobile objects, such as work instructions, strategy documents, market plans, contracts and, increasingly, communication technologies, they increase their reach and influence across space. Making different things with diverse preferences interact and relate is not easy. Law (1992) argues that different materials and actors – not just humans – have 'preferences of interaction', and patterning interactions between them is not a given. Paradoxically, the more different materials one is able to mobilise in a network, the more it becomes resistant to change:

> [What] holds society together is mostly extrasomatic. Each performative definition of what society is about, is reinforced, underlined, and stabilised, by bringing in new and non-human resources'. (Latour, 1988: 276)

So, the 'pure' social relation, if it exists at all, is not very stable (ibid.: 275). It is more difficult to hold someone accountable for a word than for a written contract. Consequently, then, the degree of stabilisation in time and space is explained by patterned ordering of heterogeneous elements. Actor-networks are related and stabilised via *associations*, the question being 'how some associations make something possible whereas others do not, how some make networks robust whereas others do not' (Hernes, 2007: 80). This means that the ordering, or stabilisation, of the social is a continuous process of transitions from the social to the material (ibid.: 83), which is by Latour called 'translation' – the work of making something represent an increasing amount of others

(Latour, 1987; 1988; 1999b). In such processes of ordering heterogeneous elements, the objects that flow, or circulate, within a network, need to be kept to a certain degree of openness for interpretation by the involved actors – 'interpretative flexibility' (Bijker and Law, 1992) – so that different actors may *inscribe* (Latour, 1987) different meanings or interests into them. If not, it will be impossible to relate the different elements of the network – often with hugely different preferences and interests – to each other. On the other hand, the degree of flexibility is limited, and it is impossible to inscribe just anything; we have to move from relativism to relationism as elements always produce some kind of resistance, often stemming from actants being interwoven in multiple relations. When the interests are too divergent, it becomes a matter of social and material negotiation, sometimes leading to compromise and at other times to disintegration, or to one actant outmanoeuvring another. The highest degree of stability (although still in process, and not *stable*), according to ANT, is achieved when patterned orders of heterogeneous elements start appearing as one object/actor, a macro-actor (e.g., an organisation) or a 'black-boxed' technology (e.g., accounting procedures), which deletes all the controversies and multiple interests that had to be negotiated to get things together. The ordering is then taken for granted, or no longer questioned or challenged, and hence *ordering* becomes (perceived as) *order*: from contingent process to necessary entity, black-boxing the insight that (in principle) things could have been otherwise. Still, even the most (materially) stabilised (social) order may fall apart, as it is only a temporal stability, and it could be said that such stability only remains intact as long as the fragility of its construction is hidden, or as long as the actor-network has the means to maintain the elements 'in place' (in more or less loyal relations to each other).

The second explanation, that resistances and limitations are experienced by actors when trying to order things into (new) patterns, is an extension of the first; heterogeneous orderings (actor-networks) are not isolated in the world, but in fact interact and intertwine with an endless number of other orderings with which they may partially overlap and partially conflict. If some degree of stability is ever achieved, this is an unavoidable aspect, as Holm (1999) has convincingly shown in his study of fisheries management. The global is made of multiple local and interconnected orderings. This point is too often simplified and misrepresented as 'structure', yet this would not be consistent with a process view. Rather, this is a case of multiple interconnected processes or actor-networks, and the interconnections are also processes that are no more or less stable than the interactive patterns making them up. This is also an issue that actor-network based analysis tends to lose; that is, in its insistence on studying only the local and situated, it unnecessarily loses sight of some of the connections to other networks in other realms. In this sense, there is a need to 'stretch out' the actor-networks of actor-network theory, without changing the principle of following the action, but ambitiously seeking to follow the action (Czarniawska, 1997) a bit farther in time and space. To change one relationship is to interfere with a number of other relationships, and, hence, the researcher of innovation processes needs not only to keep track of his 'focal actor-network', but also trace some of the related networks with which it comes to interact. We can borrow some insights from the industrial network approach here. Araujo (1998) made an early argument for merging some of the insights of actor-network theory with aspects of the industrial network approach. He advocates a network view of organisation, defined as 'a set of interlocking and shifting relations with porous and fluid boundaries'[6] (ibid.: 317), where sociotechnical networks are the units of analysis. From this perspective, the existence of boundaries becomes an empirical phenomenon; boundary practices are collective accomplishments (Orlikowski, 2002), and boundaries are dynamic (Hernes, 2004). Actors, therefore, engage in efforts to (re-)organise boundaries according to their interests in developing, stabilising and maintaining certain practices rather than others (Karlsen, 2006; Mørk and Hoholm, 2008). To take a network view does also mean that the 'environment' is understood as being 'made up of the same raw material as the organisation', namely 'multiple interactions and relationships' (Araujo, 1998: 328), or, in other words, no solid structures, only socio-material interaction processes: 'Knowing and learning are seen as collective accomplishments residing in heterogeneous networks of relationships between the social and material world, which do not respect formal organisational boundaries' (ibid.: 317).

So, knowing and learning – and I would add innovation[7] – are relational processes, to be found *between* the social and the material. They are 'implicated in everyday of collective practices' and in the 'interactional practices that relate the organisation to other actors' (ibid.). This ordering process is about shaping recursive patterns and, when interconnecting multiple such orderings, a complexity emerges that seems to privilege incremental over radical change. As Hernes (2007: 144) argues, 'reiteration is necessary in order to uphold the system of relations, and the reiteration must therefore be incremental rather than radical', because it is the incremental changes that may 'allow for movement and the possibility of discovery and connecting to other elements'. This is similar to the argument of Håkansson and Waluszewski (2001a) that, in industrial networks, innovation/change needs to be closely adapted to the existing features of implicated networks in order to be acceptable and doable, hence implying a 'conservative' logic. Yet, if this the case, what about discontinuous and more radical changes?

In reviewing the literature on (inter-) organisational innovation, Akrich et al. (2002a; b) locate a number of characteristics of innovation. However, the question of who to interact with and how to interact remains unanswered in the innovation literature. They warn us against the edifying stories employed (in retrospect) to explain the outcome of such processes, invoking, for example, absence of demand, technological difficulties or inhibitory costs as potential reasons. When innovation is in the making, these questions are all controversial (Akrich et al. 2002a: 190). We need to avoid giving retrospective explanations, using truths created by the story, and instead go closer to the actors and study innovation in the making. In order to understand innovation processes, they argue, both the diversity and complexity of the decisions to be made must be reconstructed, as well as the time and irreversibilities they create. These decisions are difficult to prioritise, because it is unknown which of them will prove to be strategically crucial. Details often end up counting and cumulatively making significant differences in achieving success or failure (Akrich et al., 2002a), in projects that may sometimes escape the hands of its inventors. It certainly is a paradox that innovation is viewed as progress by means of making decisions, and yet the outcomes of these decisions are so uncertain. Further, since innovation, by definition, is created from instability and unpredictability, no method can master innovation entirely (ibid.: 195). Hence, they argue that innovation management is about producing an environment that is supportive of the innovation. A process logic, like, for example, the one inherent in actor-network theory, tells us that we can never stop doing our ordering work. There is not a final state of order to be achieved and where we can rest. When the ordering work stops, so does the order. A 'typical' ANT story tells us about how problems and solutions are established, and how human and non-human actors are interested, enrolled and mobilised to build, support, represent and hence stabilise, the actor-network and its focal 'object' (e.g., Callon, 1986; Harrison and Laberge, 2002). To 'innovate' is to commit to a continuous process of defining, enrolling and keeping all the involved actors loyal (Latour, 1996). Consequently, to define an 'innovation' is not possible in principle, only in practice. It is continually up for negotiation, and will always risk being contested. For research then, the interesting question is how this ordering of network learning and innovation is done and maintained in practice, rather than what it is (supposed to be) in principle.

Within STS, starting out with laboratory studies, to including technology development and most recently, financial markets, the somewhat 'messy' mixing of technoscience and business requires more research, even though this impure blend of what is sacred and what is dirty has been a central characteristic of industrial practices from their very origin. In the case of this study, such 'heterogeneous engineering' (Law, 1992) entails drawing on relatively wide networks of various kinds, beyond locally situated practices, such as the globalisation of food markets. Moreover, the study seeks to show the intertwined mobilisation of technology, economy, meaning and power in order to realise industrial innovation, which is in line with Latour's (1996) study of the development of Aramis, a public transportation system in Paris, although I am tracing the network-building activities closer to actual commercial actors and end-users than became relevant in the case of Aramis. The major product of the research laboratories investigated by Latour and other anthropologists of science is texts. The major product of industrial companies, however, are (physical) objects – products – often mass produced and distributed for use by multitudes of people. Material realities and potentialities are translated to text, but, even more so, texts (ideas, strategies, patents, recipes, etc.) are translated to material products, and then to economic value. These are processes of embodying texts, and ordering assemblages of texts, materials and humans. In addition, most of the time, the outcome differs considerably from the original text (scientific text and/or business plans). Texts are not always circulated as relatively stable 'immutable and combinable mobiles' as they are in science (Latour, 1987). The creation of use, exchange and economic value is what counts (and is counted), and texts are little more than pragmatic tools on the way.

From the original cluster of anthropological and historical studies of scientific practice, STS has made several turns towards other, but still related, fields of epistemic practice. The development of technology was the first, with studies of diverse fields, from aircrafts and bicycles to bio- and nanotechnology. More recently, new directions have emerged from a number of studies of economic practices. Michel Callon's *The Laws of the Markets* (1998) was the first in a series of studies of the interaction between economic models and economic practice, and relating their findings to economic sociology and the sociology of finance (e.g., Knorr

Cetina and Preda, 2005; Barry and Slater, 2005; MacKenzie, 2006). Their contributions have been a performative view of markets, and an investigation of the constitution and role of 'calculative' agents and devices in shaping market practice. In moving towards one of the most extreme versions of economic practice, in terms of its theory-based modelling and global flows of information, the techno-economic networks of *finance* have proven to be fruitful objects for investigation.

However, large sections of economic practice are performed in a much more impure and messy mix of matters and materials. Science, technology, economic models and practices, industrial politics, professional practice, distribution and marketing practices are but a few of the elements any actor within most businesses and industries would have to take into account. In the fields of industrial sociology and anthropology, less has been done. Quite a few anthropological studies have been done on marketing and consumption practices,[8] for example, examining markets as cultures and brands as objects of cultural practice. In this study, I have found it particularly relevant to relate to the more industrially oriented works of Olsen (2000), Mattsson (2003), Cochoy (2005), Kjellberg and Helgesson (e.g., 2006; 2007b), Araujo (2007) and Brekke (2009) on the performative aspects of more 'mundane' markets, resembling studies of food distribution, telecommunications, biotechnology, electricity, car industry, etc. In their conceptual model of 'markets as constituted by practice', Kjellberg and Helgesson (2007a: 137) suggest that markets are constituted by normalising practices ('to establish normative objectives'), representational practices (to depict markets) and exchange practices ('to realise individual economic exchanges'). The links between these are conceived as translations (Latour, 1987). Markets are thus conceived of as being processes under continuous realisation in different ways, and market change/innovation is analysed as emerging actor-networks which realise their programmes via complex processes of translation, often in conflict with the interests of other (or pre-existing) actor-networks and their practices. I will get back to some specific aspects of this emergent field in the discussion in Chapter 6.

Another strand of STS studies came out of a growing concern for the *user* of technology, realising that it is difficult to fully understand technology development without including the receiving end. Not surprisingly, the findings of these studies have been that users are not only receivers; they are also participants – in more or less direct ways – in shaping the technology itself. In material semiotics (Law, 1999), it is emphasised how both the shapes and the meaning of things are mutually constituted. Within the social construction of technology (SCOT) approach (Bijker et al., 1987; Bijker, 1995), users have been included, although in a somewhat asymmetric fashion. The recent emphasis on use and users has also led to insightful studies by scholars such as Pinch and Oudshoorn (2003; 2008), who have produced more symmetrical pictures of technology development, both in relation to economy, industry, social practice, individual users, politics and the technology itself.[9]

In their review of the literature on user–technology relationships, Oudshoorn and Pinch (2008) particularly emphasise the fields of innovation studies (lead-users), sociology of technology, feminist studies, semiotic approaches (ANT), and media and cultural studies. Through empirical research identifying (some) users as innovators, von Hippel argues that many innovations come about from users identifying their own needs and trying to develop their own solutions. Thus, he came to develop the influential term 'lead-users', and has subsequently researched and experimented with ways of systematically including users in the innovation process. On the other hand, Oudshoorn and Pinch point to a study of Hoogma and Schot (2001, in Oudshoorn and Pinch, 2008: 543), who found that not all users are innovative, and hence that one needs to think through how to find a 'sensitive interactive environment for the adaptation of some radical new technologies'.

In replacing the old view of users as passive consumers, and of innovation processes as linear, the social construction/shaping of technology approaches have positioned users as *participants* in technology development, in terms of how they interpret and therefore use the technology. The boundaries between users and designers, production and consumption, are thereby blurred. The closure of the 'interpretative flexibility' of objects can be reached through several closure mechanisms, thereby stabilising the technology in a pattern of predominant meaning and use. However, designers, users and intermediaries do interact within a 'technological frame', providing institutionalised rules on how to interact with the technology. Later, the approach has taken on the challenge of mutuality – of co-construction of both users (social groups) and technologies – via the concept of 'sociotechnical ensembles'. Power is studied semiotically as relations between social groups, but due to 'the methodological priority it gives to social groups, [it] has not paid enough attention to the diversity of users, the exclusion of users, and the politics of non-use or restricted use' (ibid.: 544).

Within the more semiotically[10] inspired strands of STS, such as actor-network theory, Oudshoorn and Pinch mention the concepts 'configuring the user' and 'scripts'. In Woolgar's conceptualisation (1991), users are seen as readers of technology, and interpretative flexibility as constrained 'because the design and the production of machines entail a process of configuring the user' (in Oudshoorn and Pinch, 2008: 548); in short – the technology cannot be used in any way the users want due to the limitations built into the technology by designers. The representation of (certain kinds of) users (real or imagined) in designing and testing technology is crucial in the process of co-constructing technology and users. In other words, the process of configuring users is a matter of interaction between not-yet-settled users and not-yet-settled machines (Grint and Woolgar, 1997). Moreover, as later studies have emphasised, designers are also configured by both users and their organisations, making the process of configuration a mutual matter (Mackay et al., 2000, in Oudshoorn and Pinch, 2008: 549). In a similar fashion, 'script' denotes how designers anticipate user patterns and behaviours, and build them into the technologies, a kind of division of labour between human and machine, hence enabling and constraining social and sociotechnical relationships. Through the notion of programme/anti-programme, they show how scripts sometimes are opposed by users refusing to submit to the anticipated order of things and humans. This view challenges social constructivist approaches by giving more agency to non-humans. Actor-network approaches are criticised, however, for still giving 'more weight to the world of designers and technological objects', and therefore 'the world of users, particularly the cultural and social processes that facilitate or constrain the emergence of users' anti-programs, remains largely unexplored within actor network approaches' (Oudshoorn and Pinch, 2008: 551).

However, if designing/developing/producing a product also entails 'configuring users' in a real sense, not just 'virtual' anticipation in the design process, then how can we account for stories of mismatch between a (launched) product and (non-interested) customers/users? If the term more or less sticks to the dichotomy of illusion/reality, then the configured user would be little more than a (more or less successful) illusion, thereby losing the very real enabling, restraining and shaping interaction between the product and the user. Still, we might view configuration as the shaping of particular (and real) users, who may or may not (a) accept the terms of usage, (b) exist among the population reached and/or (c) succeed in relating to 'actual users' (Grint and Woolgar, 1997) out there.

If we now go back to Garud, who was emphasised in the innovation management section above, we have completed a circle. In drawing both on innovation management perspectives and science and technology studies, he has, together with Karnøe (Garud and Karnøe, 2001) and others (e.g., Garud and Rappa, 1994; Garud and Ahlstrom, 1997; Garud and Munir, 2008), made headway on technical and industrial entrepreneurship. In challenging the path-dependence perspective, they describe how new paths are created, and the room for agency that exists within such processes. They introduce the concept of 'mindful deviation' as a key to how entrepreneurs come to deviate from established paths and mobilise for exploration of new paths. Along the same line, Pinch (Pinch, 2001; Pinch and Trocco, 2002; Pinch and Oudshoorn, 2003) has studied the development of musical synthesisers, and convincingly demonstrated how the involvement of users and development of markets were crucial aspects for why and how some entrepreneurs succeeded with path creation while others failed.

What all of these contributions share is how the entangled elements of meanings, theories, materials, organisations, humans and technologies are carefully intertwined to hold the sociotechnical network together. Things get their value in social and economic relations, via processes of translation and of framing, in which some things are included, allowed access to a system (e.g., of a particular economy), while other things are kept on the outside or excluded. Following, changes and transformations of established systems are understood as being the effects of rivalry between competing sets of ideas/programmes and frames, and their actor-networks seeking to expand and circulate by translating and representing an increasing number of actors and elements. However, as mentioned previously, the study of science and technology on the one hand, and the study of industry and markets on the other, include very different kinds of practices and actors. Therefore, we cannot take the idea for granted that the same analytical repertoire works in both kinds of settings. This is made clear in the recent studies of industry and markets, through how they both draw on and alter the thinking of STS/ANT. I will come back to a discussion of some specific uses and alterations of actor-network theory in the discussion in Chapter 6.

Having briefly presented the empirical and theoretical starting points for my study, I will now move on to formulating the purpose and research questions to be investigated throughout.

Investigating process and controversy

If, as argued above, innovation happens across boundaries,[11] a crucial question is then how to cross or overcome such boundaries. Further, because boundaries represent *difference* of various kinds, such as different knowledge regimes, different practices, different strategies to handle uncertainty and risk, and different interests, there are usually tensions involved in the quest for the associating/aligning, translating and stabilising of elements between them. Competing programmes of action, conflicts between new and old and paradigmatic fights between knowledge regimes/discourses are but a few outcomes of such boundary-crossing efforts. Sensitivity towards conflict and controversy is strongly present in actor-network theory (Latour, 1987; 1996), which explicitly advises the researcher to trace controversies, since this is where the 'black-boxes' of relatively stable actor-networks are destabilising, hence enabling an observation of how 'new' actor-networks come about. Law (1992) describes the building of actor-networks as 'overcoming resistance' and Pickering (1995: 22) describes the production of (scientific) practice as 'a dialectic of resistance and accommodation'. Still, conflict is also present in the other perspectives outlined in the previous section. 'Friction' in the industrial network approach points to the resistances and influences that necessarily follow from any attempt at changing resource constellations. Further, the political aspect of innovating is explicitly stated both by Pavitt (2005) and Van de Ven et al. (1999). Nevertheless, I would argue that while these latter perspectives seek to include conflict and politics as part of their analysis, actor-network theory locates controversy at its centre.

There are two different viewpoints related to the actor-network understanding of how actors may be participating and entwined in a plurality of different actor-networks, performing different identities and roles in each of them. Michael (1996) emphasises the problem that sometimes occurs when an actor's identity in one setting is confronted with the identity performed in another, and how this is a potential threat to the stability of actors' identities. Araujo (1998), on the other hand, argues this is a potential trigger for learning and innovation. Both views are obviously relevant, and their points of departure are similar: The crossing of boundaries between actor-networks is likely to lead to incomprehensible performances of actors, hence possibly destabilising established order and inducing change. I also would believe that the sharper and greater number of boundaries involved, the larger potential for conflict; because identities are challenged, more interests are at stake, and it may be more difficult to stabilise (new) relations.

Given this 'conflict-oriented' view of innovation, the purpose of the study is to produce a rich description and thereby gain new understanding of *innovation processes*. How do innovation processes evolve over time? I could also refer to the *practice* of innovation, how it is done, and, in particular, how knowledge and technology are being developed and commercialised. I seek to produce insights related to this as an empirical problem, which is recognised as a problem by most of the practitioners I have been talking to, and where the available knowledge is scarce. However, it is also a theoretical problem, and I seek to contribute to perspectives on innovation that we could call process-based, relational/ interactive and situated.

In order to identify a suitable case for the task, I relate to Van de Ven et al.'s (1999) definition of a 'generic innovation journey', emphasising innovation processes that are purposeful for developing a novel idea, yet constitute substantial uncertainty regarding the market, technology and organisation, a collective effort over time and require greater resources than those possessed by the people who undertake the efforts (ibid.: 22). In connection with the setting where my project is based, the Centre for Cooperative Research at BI Norwegian School of Management, I spanned the partnering companies for interesting and ongoing innovation processes. Tine BA, an agro-food cooperative with its core business in the dairy sector, turned out to have a combination of new corporate innovation strategies, a large R&D department and a number of interesting ongoing innovation projects crossing the industrial boundaries between the agricultural and biomarine industries. In addition, in order to enable looking into the 'real' struggles of innovation – with all its contingencies, uncertainties, controversies and heterogeneous elements combined – I assumed that 'radical' and boundary-crossing projects – that is, the projects with the greatest gap between the established activities and the innovation projects – would be more open to investigation and display more of the innovation dynamics at play. Very little in these agro-marine innovation projects had reached any degree of 'stability'; the final shape and destiny of the innovations were still undetermined.

It is impossible to explain what happens in innovation processes in individual, deterministic or linear ways. These are highly interactive processes, involving a number of both human and non-human elements, where the outcome – on almost any parameter – is not given at the outset. It is also impossible to simply transfer knowledge, without also transforming (or destroying) it. Hence, the research questions I have pursued in this study are the following:

- *How do innovation processes evolve over time?* This is a very open and explorative question, corresponding with my ethnographic approach and my basic interest in innovation as temporally emergent and culturally situated.

- *How is knowledge translated, transformed and combined in processes of innovation?* And what kinds of knowledge are involved? The processes of translating, combining and transforming knowledge necessary for realising innovations is still not well understood in industrial settings.
- *What are the contrary forces of innovation processes?* Beginning with acknowledging the presence of controversies in innovation, I want to understand more about what dynamics produce and fuel the inherent tensions of innovation processes.

These questions have grown out of a circular process of going back and forth between the literature and the field. They are designed to capture both how actor-networks are built for innovation, how knowledge and innovations are developed and how the broader settings of the process influence the actual process under investigation. Hopefully, this may contribute to innovation studies by shedding some light on the question of how knowledge, which is developed in certain contexts and embodied in certain objects and practices, is rendered mobile, translated, combined and made stable in new settings, objects and practices underpinning innovative ventures. 'Knowledge' is here only analysed in terms of how it is materialised in technologies and work practices. Thus, the focus is on *knowing*, or the doing of knowledge – knowledge as a performative construct inseparable from the historical, social and technological setting in which it is embedded (Law, 1994; Araujo, 1998; Gherardi et al., 1998; Gherardi and Nicolini, 2002). Industrial innovation refers to the process of developing, producing and commercialising new objects through recombining, transforming and translating knowledge and technology. Hence, innovation is about the whole process, from the inception of an idea until its eventual implementation/commercialisation (or failure).

The structure of the book

The next chapter will account for the use of organisational ethnography, and a discussion of the challenges of doing ethnography in studies of innovation processes. The third chapter is a brief presentation of the case I have been investigating, and an outline of the wider industrial and political setting of which it was a part. On the one hand, there is the emergence of 'blue-green' innovation policy and how actors in science and industry mobilise this institutional discourse to support their local innovation efforts. One the other hand, there is the development and combination of two innovation programmes, both seeking to develop new 'value-added' products from fish. The fourth chapter provides an in-depth description of the emerging innovation process, and the struggles to move, combine and stabilise knowledge and technology across boundaries. It is a relatively detailed description, being an example of how micro-practices both create and enact macro-practices: first, because macro-practices are built by interconnecting micro-practices and, second, because the emerging macro-practice and the emerging micro-practice mutually shape and influence each other.

In the fifth chapter, I outline an analysis scheme for studying and analysing (industrial) innovation processes that, in a simplified way, seem to capture some core aspects of this innovation process. Based on my fieldwork and the methodological-analytical basis accounted for in the first chapter, I suggest that innovation processes may fruitfully be conceptualised as a dual process of mobilisation and exploration, of facilitating and attracting resources for innovation, and of actually formulating and testing propositions about reality. Thus, I arrive at a bipolar model, in which the particular dynamics between the two poles of a concrete innovation process become a central part of explaining the case. In the sixth chapter, I draw out some of the theoretical implications in discussing how my study may contribute to and complement theories of innovation processes, using the outlined analytic scheme to interpret the case. Finally, the seventh chapter concludes the study by summing up my findings and pointing at some areas for further research.

2
Constructing Ethnography

In this chapter, I will discuss methodological concerns regarding the (inter-) organisational ethnography of innovation processes. There are two initial premises that have influenced my research aims, and my theoretical and empirical choices in this study. First, in the literature there has been a clear call for 'process studies'[1] of innovation and organisation (Tsoukas and Chia, 2002; Van de Ven and Poole, 2005; Hernes, 2007), which involves making change – or process – the point of departure, and thereby placing *stabilisation* of innovation and organisation at the centre of attention. There is a need to improve our understanding of the mechanisms and dynamics of how organisation and innovations unfold in practice, or of how they come into being. Therefore, I searched for interesting places to study innovation processes in real time through (inter-) organisational ethnography. Second, when I was granted the opportunity to study innovation processes from the Centre for Cooperative Studies at the Norwegian School of Management BI, the main agricultural food cooperatives in Norway – who were sponsors of the centre – became the most relevant empirical fields of study[2] – that is, if I could find relevant and interesting cases there.

To study innovation processes is to study an emerging object or practice from the inception of an idea to its realisation (or failure). Further, it involves studying the interactional processes of the involved actors, whether they be scientists, engineers, managers, marketing and production staff or customers, governments and finance institutions, not to mention the non-human actors, such as technologies, texts and buildings. Studying innovation is about observing and accounting for an object that is also a heterogeneous network: to investigate how ideas, knowledge and meaning gradually get transformed and embodied, and thereby making the innovation more real. This is why studies of innovation processes lend themselves towards the situated, relational and narrative approach of actor-network theory.

There are many problems with researching innovation. First, accounting for outcomes of social activity demands analytic tools that are able to include both humans and non-humans, as the social is socio-materially constituted. Second, accounting for interaction processes, including how things get stabilised and destabilised is a very different task from the more common social science methodology of measuring input and output factors to prove causality and significant relationships, or to map elements of a social realm as if they were static, stable and generalisable. Third, if only 'hard facts' are accounted for, it would be impossible to tell stories about the ordering of the social: we need to include the intentions, strategies and compromises that are made, and how actors inscribe meaning into their materials and activities. While Gupta et al. (2007) with good reason call for more complex and comprehensive studies of innovation processes, I do not agree with them that such studies should be 'multi-level'. The argument against multi-level analysis is important to this study, as well as not 'settling the question of scale in advance' (Hernes, 2007: 74). When the social researcher starts following the action, or the 'connections and associations made between heterogeneous actors' (ibid.), the term 'context', and the distinction between micro- and macro-levels are no longer relevant as analytical concepts. Context, if anything, becomes an empirical phenomenon of how the actors draw boundaries and 'frame' their activities.

How does this inform empirical research? At best, to practice ANT is to build research on careful empirical observations, and acknowledge some central principles. First, it is 'important not to start out assuming whatever we wish to explain' (Law, 1992: 380); instead, Law suggests to 'start out with interaction, and assume that interaction is all there is', rather than starting out with some abstract overarching concept, like class or structure. And from there ask questions about how some interactions manage to reproduce themselves into more stable orderings. How do they 'overcome resistance and become macro-social' (Law, 1992: 380)? Latour (1987) outlines a set of methodological rules, and a set of principles for analysis emphasising the overarching ideas of symmetry, agnosticism and free association. These principles are also discussed by Callon (1986), who makes a rather sharp critique of sociological theorists for their asymmetric analyses. This becomes visible in the paradox that sociologists act as if the *agnosticism* they apply to natural science and technology, which allows for a plurality of descriptions, is not applicable towards society as well: nature is uncertain but society is not. In this way they remove their own knowledge from public discussion (ibid.: 197), which means that society is always given the last word. Callon's argument is that both the social and the natural are equally uncertain and ambiguous, and hence disputable. He therefore generalises the principle of agnosticism, or of analytic impartiality, stating that one should not a priori privilege humans above non-humans. Thus, sociology has no solid foundations, and 'is as debatable as the knowledge and objects which it accounts for' (ibid.: 199). The heritage from semiotics and post-structuralism become visible in Callon's description of scientific and technological innovations as 'dramatic stories' in which both the identity and importance of actors are at issue (ibid.). ANT can fundamentally be described as a way of writing stories about the social, based on a set of research principles and methods. Furthermore, the choice of which methods and what analytical repertoire to use is up to the researcher according to what seems best suited to the actual task. Yet, there should be one single coherent repertoire to secure the *symmetrical* analysis of all actors in scope, and it is then the researcher's task to convince colleagues that the right choice has been made (ibid.: 200). However, this repertoire should not include pre-established analytical grids or categories; rather there should be an attempt to follow the actors to understand how they build and explain their world (ibid.: 201). Hence the principle of 'free association' is introduced, emphasising idealist methods such as ethnomethodology, phenomenology, discourse analysis, etc.

However, some authors have also posed critical questions of the actor-network theorists' approach to ethnography. In summing up some of these critiques, Vickers and Fox (2005) mention some arguments about representation. ANT has tended to focus on elites, a narrow set of actors: despite its claim for symmetrical analysis it still tends to represent some humans more than others. ANT has also been accused of refusing to engage in ethical debate, taking a relativist standpoint. However, Vickers and Fox suggests that there is nothing innately relativist in ANT as such, though this still might be the case in some of the ANT studies that have been done. Some argue that ANT is marked by 'analytical decontextualism', that the 'wider social context is often let out of the analysis', and even that ANT researchers only pick out the bits from their ethnographies which corroborate their ANT points. Vickers and Fox's solutions to this issue are to focus more on non-elite people, and on processes of counter-enrolment/resistance. They further accept that ANT offers a kind of decontextualisation, like other ethnographic approaches, the context in sight is dependent on the viewpoint of the studied actors and those of the observer. However, in line with the decontextualism argument, I find that there is good reason to seek to 'stretch' actor-network ethnographies a bit beyond what has commonly been done in this field of research; tracing the network-building activities beyond local sites and projects, and connecting the narrative to some of the other narratives with which it interacts in practice. My descriptions in Chapters 3 and 4, and some of the subsequent discussions in Chapter 6 have sought to deal with this.

Since the late 1970s, ethnography has become a common approach to studying knowledge production within science and technology studies (Hess, 1992; Law, 2004). According to Law, this is because it reveals 'the relative messiness of practice', helping us to 'understand the often ragged ways in which knowledge is produced in research' (Law, 2004: 18–19). Acknowledging the need for investigating *practices* of industrial innovation, including its situated and contingent character, makes a good case for ethnography. If a central issue, perhaps *the* central issue, in industrial innovation is knowledge (both when it is available and when it is lacking), then it is reasonable to think that similar methods could be fruitfully applied to understand how knowledge is produced in such settings. This is consistent with Hess's (1992: 14) call for decentring the laboratory in STS studies, as knowledge production is distributed across a number of different settings and practices. The main differences from most ethnographies of science and technology are perhaps, first of all, the distributedness and heterogeneity of industrial practices, often involving more heterogeneous constellations of actors. This certainly includes scientists and technologists, but also marketing and sales personnel, business managers, logistics and distribution actors, politicians, bureaucrats, investors and customers. This means that action is found in many places, often at the same time. It also means that the ethnographer is challenged to engage in observing and understanding very different kinds of practices, though s/he does not always have the privilege of focusing on just one of them, and is certainly not able to capture 'everything' relevant to the process. In this sense, this study has not only decentred the laboratory, but also business management, which has been put at the centre of a great deal of business research. Corporate board and management in the organisations studied are just a few of the many places and practices I had to visit; in fact, most of my time in the field was spent with 'operational' project participants, such as middle and project managers, scientists, technologists and marketers.

My aim has been to 'follow the actor' (Latour, 1987), wherever the action happened to unfold, to understand the processes and practices of industrial innovation in a broader sense; to avoid the managerialist and the technological determinist pitfalls. Secondly, industrial innovation involves a set of epistemic and economic practices which have not yet been studied extensively within STS.[3] These are practices with different aims, frames and evaluation criteria; the stabilisation and evaluation of knowledge is performed less according to scientific norms of knowledge production than economic norms of profitability, return on investments, etc. Hence, transformations of scientific knowledge into industrial and economic practices are often uncertain processes demanding considerable time and resources (Håkansson and Waluszewski, 2007a).

Advantages of this methodology are the opportunities following social (and socio-material) processes and practices *as they evolve*. Ethnography may be said to be about participant observation, with the ethnographer,

> participating, overtly or covertly, in people's daily lives for an extended period of time, watching what happens, listening to what is said, asking questions – in fact, collecting whatever data are available to throw light on the issues that are the focus of the research. (Hammersley and Atkinson, 1995: 1)

Emphasis is put on interaction between the ethnographer and actors in the field, hence the argument that ethnographic fieldwork has a dialogical nature (Hess, 1992). Such real-time studies have more power to elucidate the uncertainties and contingencies the actors experience in the course of deciding and acting. This is not a matter of constructing an objective truth of what happened; rather, it is an attempt at reconstructing the actors' experiences, interpretations and actions in the face of the 'opportunities' and 'uncertainties' of innovation. In other words, I wanted to reconstruct some of the difficulties, controversies and choices that the involved actors had to handle, and avoid post-hoc rationalisation. Latour (1987: 258) underscores the need for studying knowledge production *in action*, in order to 'either arrive before the facts and machines are blackboxed or... follow the controversies that reopen them', thereby looking for the transformations the innovation goes through during the process of its realisation. In addition to the advantage of 'real-time' studies of contingent processes, and the provision of 'thick descriptions' (Geertz, 1973) of the empirical field, ethnography may produce both deep insights and increased variety of interpretations. The ethnographer is not granted sovereignty over interpreting and theorising the case. If research is viewed as being an ever-evolving discussion, this should be viewed as a great advantage, and if the analysis of social issues is the researcher's 'constructions of other people's constructions of what they and their compatriots are up to' (ibid.: 9), his/her discussion, reinterpretation and comparison with other cases should be viewed as a necessity. However, in my process of investigation, I found it necessary to include a set of events that happened before I entered the setting. In these cases I have had to rely solely on documentation and interviews with participants in these events. Even if most of these events happened during the reasonably recent past (from less than a year to three years before), and I cross-checked information with different actors, I cannot rule out the possibility that I missed important aspects of these events. Still, the aim has not been to find the ultimate truth about these events, but to reconstruct and re-present the challenges the actors were facing as much as possible. Thus, in terms of the process of gathering data, there are two different methodological challenges during the real time and more historical parts of this study: the distributedness of (inter-) organisational ethnography in real time, and the danger of actors' post-hoc rationalisation of past events. I have sought to handle these challenges by cross-checking historical materials, and through keeping in regular contact with key informants in order to catch up with recent events when they were still 'fresh' and their meaning had not yet been collectively stabilised in the organisation. Cox and Hassard (2007) review and discuss retrospective methods in organisational research, and warn against positivist and interpretivist positions that assume the past may be controlled or distinguished from the present in retrospective research. Instead, they propose a position of re-presentation, in which the present is not understood as being independent of the past, but rather that the past, present and future are co-constituted both in the negotiation of meaning in organisations, and during the writing of the researcher. In this sense, the stabilisation of history is an interesting topic in itself – of how actors delete and/or reinterpret aspects of their past in their ongoing processes of realising their present projects and identities. The researcher also has to acknowledge that his story is just one of a number of potential versions, account transparently for how this came to be and make an argument for its value. The resulting time and shape of the processes described are therefore both products of the participants' negotiations, and of the researcher's purpose, questions and fieldwork (who I talked to, what documents I got access to, when and where I got to observe the ongoing processes, etc).

According to Geertz, interpretive approaches to social research 'tend to resist...conceptual articulation and thus to escape systematic modes of assessment' (1973: 24). But, as 'social actions are comments on more than themselves', there is 'no reason why the conceptual structure of a cultural interpretation should be any less formulable...than that of, say, a biological observation' (ibid.: 23–24). In this book, I have written about the empirical research in Chapters 3 and 4 in a relatively descriptive way. This means that, to a large extent, I have sought to provide 'thick descriptions' of the case, and suspend most of my reflections and theorising discussions until the subsequent, Chapters 5 and 6. It does not mean, however, that Chapters 3 and 4 are 'objective' in the sense of 'not analysed'. Of course, during the entire process of choosing case and questions, of observing and talking to actors in the field, and especially choosing what to include and how, this written account has been highly influenced both by the theoretical fields mentioned in Chapter 1 and my own reflection and interpretation along the way. By displaying the case study in this way, in a style relatively close to what Van Maanen (1988) has called a 'realist tale', to not include my own person or much of my reflection as part of the empirical account can easily be seen to produce an illusion of objective reality. And this illusion is possibly strengthened by the extensive use of quotes from interviews (and to some extent documents), at the expense of excerpts from my own field notes, even if much of my understanding and interpretation of the case has been shaped by actually being in the field. Much recent ethnography has embraced styles closer to 'confessional' or even 'impressionist' tales (ibid.), in various ways including the ethnographer with her preconceptions and influences on the empirical situation (as 'participant' both when being in the field and when reconstructing the story in text), in particular in studies sensitive to power relations, such as critical management studies and feminist studies. On the other hand, such reflexive modes of writing ethnography have sometimes come to be rather 'researcher centric' – that is, they deal more with the researcher's own process than the process under investigation. For reasons of clarity and focusing on the research questions, I have therefore chosen – in writing – to do three things separately: first, a transparent account of my own research process, including theoretical starting points and methodological process (Chapters 1 and 2); second, a rich case description with an extended introduction to situate the case (Chapters 3 and 4); and, third, I have tried to be clear and bold in fleshing out my analysis and arguments when theorising from the case study (Chapters 5 and 6).

Digging into potential cases

Empirically, I started out early in the project with mapping innovation strategies and activities at Tine (the former 'Norwegian Dairies') and Gilde ('Norwegian Meat'), both farmer-owned and highly industrial-ised cooperatives. I met up with directors and middle managers with a stake in these companies' innovation strategies to identify and discuss ongoing innovation processes in the different settings. In particular, I was looking for cases and processes where the organisation would have to 'stretch' their knowledge and resources beyond what they needed in their well-established core activities. This would mean a need for them to learn about how to recombine their expertise and resources with others' knowledge and resources. Moreover, they would not be able to rely too strongly on their established routines for research, product development and marketing. In following such processes, it was pos-sible to get fairly good data not obscuring the uncertainties and contin-gencies of innovation-in-practice. I soon came in contact with people like Mæhle (corporate director, Tine), Hovland (managing director, Tine Ingredients) and Skjervold (innovation manager, Tine R&D), who were all involved in shaping Tine's biomarine strategies and activities, and I understood that there were a set of projects here in which the com-pany's competence and resources would be challenged. For Tine, sev-eral of these projects represented relatively 'radical' innovations, both from a technical and a market perspective (see Chapter 3 for an outline of Tine's biomarine innovation strategy and the related projects). They started out with the assumption that their established expertise, tech-nology and infrastructure would be transferrable to the fish and other biomarine industries, but without much knowledge of what it would take to get there.

At the same time, I participated in a number of the meetings of a small group called 'Jarl S. Berg's experimental factories' (JSB); an initia-tive for enhancing creativity and innovation across professions in Tine R&D at Kalbakken (Oslo).[4] They had work meetings once a week and this became a valuable source of information and insight into this part of the organisation. Research manager Eirik Selmer-Olsen was leading the group at the time, and he generously opened doors for me so that I could gain access to the R&D milieu at Kalbakken. Moreover, several people who were or had been involved with the Neptun and the Umi No Kami projects were members of the group: Lars Petter Swensen, Elin Simonstad Valle, Svein Erik Hilsen and Even Manseth. This helped me understand the large and complex R&D organisation in Tine, getting to know some of the 'blue-green' thinking on the research side, and build relationships with people who became important informants to my study. In this phase, I also considered the JSB group to be an interesting research object in itself. However, as I enquired into some of the bioma-rine projects, I found these to be better suited; more concrete, higher priority in the organisation and more urgently having to stretch far beyond the Tine organisation, both in relation to industrial and market practices and knowledge.

In particular, two projects seemed relevant for my purposes: they were ongoing when I arrived, they entailed significant challenges on issues of organisation, technology and market, they brought hopes for syner-gies with Tine's existing business and they had high priority on the top management levels in Tine. During my first year of the study, therefore, I traced both the Maritex and the Umi No Kami projects. Maritex was a fish oil and by-products factory that Tine at the time owned, together with Aarhus Olie (Danish producer of vegetable oils); Umi No Kami was the project of fermenting fish into a salami-like product. I was granted good access to both projects, and started investigating the project his-tories, talking to project participants and visiting the project partners (Bremnes Seashore in Bømlo and the Maritex factory in Sortland), as well as talking to the management there. As both projects were based in Tine Ingredients, a small department of around 15 people, and with strong connections to Tine R&D, I could combine investigations of both projects at the same time.

The projects increasingly were based on very different business logics (omega 3 as nutritional ingredient versus Salma as high-end consumer brand concept), and, as progress was made, the Umi No Kami/Salma project and its set of innovation processes manifested all the aspects I was looking for in my study:

- A novel product technology based on a combination of marine and agricultural technologies and practices.
- A resulting product that was experienced as radically new and hav-ing an uncertain categorisation within existing food markets.
- A networked organisation of the project, involving actors from the fish industry, the agro-food industry and from academia.

All these aspects suggested a need for extensive learning and knowledge production across professions, organisations and industries. Moreover, the degree of uncertainty involved, and the amount of time and resources invested, meant that I had the chance to follow – partly in real time – a fascinating innovation story with very explicit examples of what is often called 'radical innovation'.

Doing organisational ethnography

Gathering materials

In line with ethnographic methods, the fieldwork of this study consists of gathering a highly heterogeneous mix of research materials, using participant observation, informal conversations, formal interviews and document analysis. During the spring and autumn of 2004, I spent a lot of time at Tine R&D, both with the JSB group, and talking to various people in the organisation. I also began conversations with Hovland, Mogård and Kiland at Tine Ingredients, and, in June 2004, I went with Kiland and Swensen to Bremnes for the first large-scale production of the salmon salami there. I also joined Swensen on a new trip to Bremnes in September 2004, and finally visited Bremnes in August 2006 for a second round of interviews with the management of Bremnes Seashore and Salmon Brands. During the autumn of 2004, I increased contact with the social milieu at Tine Ingredients, and in October 2004 I joined Kiland and a number of other Tine representatives at the SIAL food fair in Paris, where Salma Cured was presented, together with the Jarlsberg cheese and other Tine products, to a large number of international business actors. In November 2004, I went with Bente Mogård and Roger Hem to Maritex in Sortland, and had the opportunity to discuss the Tine–Maritex relationship with the management team there. From January through April 2005, I borrowed a desk in the open-plan offices of Tine Ingredients. I spent several days a week there to observe their work practices, strengthen my informal dialogue with central actors in Tine's biomarine projects, conduct a first round of interviews and go systematically through available project documentation in project and individual archives. Several of these people became key informants in my fieldwork, people who I met relatively often; they were in various ways involved in the processes I wanted to study, and willing to openly share their views and experiences with me. With regard to my status and role during the fieldwork, as a newcomer to the organisations I was allowed the role of 'acceptable incompetent' (Hammersley and Atkinson, 1995: 103), thereby having the opportunity to ask 'silly' questions about things 'insiders' take for granted. Related to Junker's typology of social roles in the field, I was perhaps closest to the 'observer as participant' role, meaning that I did not take part in actually doing any of the activities I studied, and I did not have any tasks or responsibilities within the organisations. At the same time, everyone knew that I was a researcher, and I spent a great deal of time talking and interacting with the actors. This is 'participant observation' (Hess, 1992), not to just observe behaviour, but also to engage in dialogue. As with many other ethnographers, the informal conversations at the desk, by the coffee machine and over lunch provided me both with valuable information and with a deepened understanding of the practices of the organisation. I sought to keep in regular contact with people at Tine R&D also during this period, although not as often as the year before. More of the action in the Umi No Kami/Salma project had moved to Tine Ingredients, Bremnes Seashore and various customer locations. I also experienced that my relationships with the people in R&D made it easier to contact them and catch up with recent events.

In total, I have conducted 35 formal interviews, 21 of which were taped and fully transcribed. The other interviews were either done via telephone and in more spontaneous situations, in which it was not appropriate or practical to tape the session. Instead I took notes, and transcribed these interviews and meetings as soon as possible after the event. Right after all interviews, I also took notes on my own reflections and interpretations of the situation; how the interviewees responded to my questions, what they emphasised and their suggestions about where to trace the processes under investigation. A few key persons were interviewed twice, both to complement me on particular issues and to capture aspects of the process over time. I did not use a structured interview guide during any of the interviews; instead I brought lists of topics that I wanted to cover during the conversation. This is, according to Hammersley and Atkinson (1995: 154), perhaps the clearest difference between the way survey interviewers and ethnographers structure interviews: between 'standardised' and 'reflexive' interviewing. This is not a matter of unstructured versus structured; rather, the ethnographer structures the interview – together with the interviewee – as conversations, where the order and mode of questions may shift as the conversations evolve; non-directive, directive and even confrontational.

In addition to these formal interviews, my understanding of what was going on, including emerging innovation strategies, work practices and power relations, has to a large extent been shaped by the informal interaction over time with the various participants in the studied processes. Meeting up with individuals and groups during their daily work activities, sitting at a desk in their open-plan offices and travelling with them to partner and customer meetings produced a large number of interesting observations and informal conversations. For practical reasons, I did not have the opportunity to balance my time spent 'in the field' evenly throughout the investigated process or across the places where things happened. First, this was because parts of the processes had already taken place. The Neptun project was finished, as well as the first phase of the Umi No Kami project. For these parts of the story, I have had to rely on a combination of interviews, informal conversations and document analysis. I have therefore also included interviews with key personnel from these early phases who were no longer participants in the projects. Second, the real-time processes under investigation were unpredictable and complex. Sometimes meetings and discussions had taken place at short notice in times and places where I was not present. Even if I experienced an open attitude to my presence, it did not always mean that I was invited to business meetings or other events of 'potential impact'. At other times things happened several places at the same time, like when the people at Bremnes struggled to improve and stabilise their production routines together with people from Tine R&D, the marketing people worked with adapting their strategy towards potential customers, and the management worked on renegotiating agreements between the involved parties at the same time.

During my fieldwork in Tine, I had open access to project archives from the Umi No Kami and Salma projects, and also the Maritex project, including project descriptions, strategy documents, meeting minutes, market research, board documentation and even some email correspondence between involved parties. During my visits to Tine Ingredients, I spent much of the time I had between conversations and meetings systematically going through all available documentation, taking notes on information I found useful for my study. In spite of this openness and the access I was granted to many of the settings where these processes evolved, there was one aspect I could not satisfactorily cover on my own. I did not have access – or capacity – to follow the processes that took place at the top management of Tine and the interaction between the top management of Tine and Bremnes Seashore on an ongoing basis. I could partially compensate for some of this by maintaining a continuous dialogue on these matters with the project participants who would maintain a dialogue with the top management, such as Hovland and Kiland, and I also have had several meetings with Mæhle (corporate director, Tine) and Tryggestad (board member, Tine). In addition, I have conducted formal interviews with Refsholt (CEO) and Mæhle from Tine, and with Morlandstø (managing director, processing) and Svendsen Jr (chairman/owner) from Bremnes Seashore.

Still, from a process studies perspective this only took me so far. However, two of my colleagues at the Centre for Cooperative Research (BI Norwegian School of Management), researcher Margrethe Schøning and Professor Morten Huse, did a study of Tine's board meetings in 2004 based on participant observation and interviews. They were helpful in taking particular notice of the board's discussion of the biomarine cases in the board meetings, as well as touching upon biomarine issues in their interviews with board members, and Tine accepted that I gave aggregated presentations of the main aspects from these board meetings. This has been particularly useful in my description of Tine's biomarine innovation strategy (Chapter 3) and, to some extent, in descriptions of Tine's decision to change the direction of the Umi No Kami/Salma project during the later phases (Chapter 4). Finally, I have on a few occasions had the opportunity to present my (preliminary) interpretations of the innovation process back to project participants and well-informed groups of people: in meetings with strategic staff, middle managers and project participants from Tine and Salmon Brands, in seminars with the Tine board members present and by getting project participants to read and give feedback on the different papers describing and analysing the case study.

Before moving on to my own process of writing ethnography, another issue needs to be accounted for. In my aim to provide an in-depth description of the processes going on in the case I was researching, it became natural to keep the account transparent with regard to the participants in the study. No one has been anonymised,[5] and all sources of direct quotes are identified in the text. This should not be problematic, first because all participants have been clearly informed of my role and intentions as researcher, and the study was initiated in close dialogue with top and middle management in Tine. Second, I sought to treat all participants with respect and have seen no reason to pass judgement on their actions. My interpretations have also been presented and discussed with central actors in this study on several occasions. Third, all sensitive topics related to strategic issues (e.g., competition) will have already been made public by the actors themselves at the time this book is published. Fourth, and most importantly, I take full responsibility for the entire reconstruction of the involved processes and actors; a reconstruction with specific purposes and methods leading to one out of a multitude of potential versions.

Reorganising materials: writing ethnography

To be sure, in ethnographic research, observation does not precede analysis as they are better depicted as constituting an intertwined process, a 'dance' between observing, talking, reading, thinking and writing. However, as many anthropologists have noted, perhaps the most demanding task of ethnography is *writing* (Geertz, 1973), so methodological resources are also needed for the textual treatment of the often massive amount of field materials (notes, interviews, documents, artefacts, pictures, videos, etc.). Writing is a process of ordering these materials into a meaningful text, that is, a text that provides new insights into the particulars of the investigated setting, as well as what can be learned from this in dialogue with other studies of similar phenomena. Thus, how can we account for practice-based studies of innovation processes? An ethnographic research strategy in inter-organisational settings tends to produce an enormous amount of detail, which is incomprehensible without some framework through which the story can be reformulated and analysed. An 'ordering strategy' was needed for handling the complex data (or 'capta' as suggested by Hernes, 2007).

During the period from August 2005 to August 2006, remaining interviews were done, while simultaneously continuing transcription work and analysis of field materials. All interviews and field notes were analysed, using the NVivo7 software as a 'cut and paste' tool[6] when identifying and sorting the materials in themes and events, both in looking for themes 'growing out of' the texts (including my experiences in the field) and themes related to research questions and theory. In this circular process of observing, analysing and reading theory, I had to reanalyse and reorganise the materials several times before I was confident that I had a set of themes and stories that would help me answer the research questions in a constructive way. This was also related to my attempts at structuring a narrative and theorising from the empirical materials. Many ideas about how the story should be told, and how theoretical discussions and contributions should be framed, were tested in writing and discussion with colleagues, before this book took shape. Should the chronology of the investigated events, the themes I want to emphasise, or even my own process of investigation be the underlying principle for presenting (reconstructing) the empirical stories in text? I ended up deciding on a relatively chronological and detailed description of the case (Chapter 4), preceded by an introductory chapter, which situates the case study in a network of interconnected processes (Chapter 3).

In line with Geertz's (1973) concept of 'thick description', and Hess's (1992) suggestion to include more of the field materials in the account to avoid finite interpretations from a 'superior' ethnographer, I have used field materials extensively. However, it has been a lot easier to be explicit about using interview materials and documents than using field notes of my own observations and experiences in the field. This has partly to do with the danger of ending up with a researcher-centric 'confessionalist' tale, and partly with the sense of certainty: this is really what the informants said, whether the reader agrees with my interpretation of it or not. Nevertheless, although the aim here is to make the case study both more transparent and convincing, I still have to carry the responsibility for the questions I asked, what quotes I included in this text and how I combined quotes and my own thinking.

Having finalised a version of the case study that displayed the contingencies, politics and knowledge-producing aspects of the story, I made another attempt at shaping a theoretical discussion of innovation processes. From my interpretation and reconstruction of the empirical story, which was clearly inspired by related research, a dual – and contrary – dynamic seemed particularly noticeable, between the activities and processes related to mobilising decisions and resources for innovation on the one hand, and activities of actually developing and testing the innovation on the other. An analytic scheme (Chapter 5) was then developed to amplify this dynamic and thereby guide and focus the theorising process. Based on (a) an actor-network theory-inspired research design, (b) an in-depth empirical description of complex innovation processes and (c) this amplified and simplified analytic scheme, I framed a theorising discussion (Chapter 6) which aimed to draw on, challenge and complement existing theory on innovation processes. In line with Weick (1989), the theorising in this study has been a process involving creative imagination, and simultaneous parallel processing of research materials, theory and my own interpretations. With regard to the outcome of the study, I have aimed at contributing to the field of innovation studies, first, by providing rich insights into an under-researched phenomenon and, second, by engaging in dialogue with related literature on conceptualising key characteristics of innovation processes, and discussing the relationships between them (see, e.g., Weick, 1989; Walsham, 1995; Hammersley and Atkinson, 1995; Law, 1994, for discussions of generalisations from interpretive research). The theorising of this study is therefore best understood within the category of conceptual generalisation. I have used my empirical account and the following analytic scheme to suggest the theoretical implications for innovation processes, while challenging and complementing previous studies. Still, the degree to which the insights from my study of a particular set of innovation processes within the food industry can be fruitfully applied to innovation processes in other settings remains a question for future research.

All in all, this study has been an explorative and inductive search for more precise process-based understanding of the organising, or the practising, of industrial innovation. Such understanding has been sought out, first in making an effort towards methodological rigour and continuous reflection on the limitations of the study in terms of the above-mentioned problems with time, timing and place. Second, the study has been conducted with a continuous cycle of intensive interaction between existing theoretical conceptions of innovation processes and the gathering of empirical materials. Third, versions of the story and their theoretical implications have been discussed with research colleagues on a large number of occasions, such as department seminars at the Norwegian School of Management BI, and international research conferences. When finalising this book, I have put particular emphasis on keeping the empirical account rich and transparent; enabling the competent reader to find more to the story in the empirical descriptions than that which I have exploited in the subsequent discussions.

There are some limitations to this study. Observing at multiple sites, not knowing where (influential) decisions are made and having limited access to the field have been the most pressing issues during this research. In addition, I had to use historical materials in order to trace parts of the process that happened before I started the field study. Related to theory, this study is limited to just one case study, hence limiting the discussion of generalisability of my findings. This is partly compensated for by comparing with previous literature, but the relevance of the analytic scheme in Chapter 5 and the theorising in the subsequent chapters for other cases of industrial innovation remains unexplored: in what settings and under what conditions will the model be useful? How should it be tuned for (1) handling a wider set of innovation processes, or (2) giving a more precise and comprehensive understanding of particular innovation processes? In addition, I cannot exclude the influence of the specific setting of the case study for developing the analytic scheme and the subsequent theorising. Nevertheless, I presume that the insights of this study have at least some relevance to other industrial settings, as many similar elements are likely to be present.

Part II

3
Introducing the Case Study

This chapter is an introduction to the cases that were investigated in this study and the setting in which they emerged. On the one hand, I observed the emergence of 'blue-green' innovation policy[1] and how actors in science and industry mobilised this institutional discourse to support their local innovation efforts. On the other hand, I observed the development and combination of two innovation programmes, both seeking to develop new 'value-added' products from fish. One programme explored the innovative use of fermentation technology for curing fish, while the other was built on novel technology for processing 'super-fresh' salmon. In *principle*, these could be viewed as separate projects that did not have much to do with each other. Still, in *practice* these projects became partly integrated and partly competitive, and some of the most interesting aspects of this study are found in the interaction between these two projects, and their subsequent combination into a single brand, organised in a joint venture.

Emerging innovations are at the centre of the narrative. Their relations to other actants and actor-networks – organisations, humans, technologies, supermarkets, etc. – are traced in order to understand more about the contingencies inherent in the process. Controversies, ambiguities and coincidences say a great deal about the potential direction of the process along the way, and history, established technologies, competences and relationships, and dominant economic systems speak to stability and friction – or why some things are difficult to stabilise, and how these processes have unintended consequences.

The case might be seen as an early example of the industrialisation of aquaculture, or signs of an emerging transformation of the industry, even though this particular story is situated in a local and historically contingent setting. The resulting products are testament to such a transformation. First, there is the industrial texture in the salmon salami, blended and cured, which presents itself as 'pure' – a perfect blend of ingredients, and representative of an industrialised food item. Second, there is the scientific/industrial development of processing technologies for fresh salmon, and the following 'branding' of a sophisticated version of a generic product: salmon loins. Third, there is the industrial economic logic, leading to the development and recombination of production technologies, packaging and design practices, logistics systems and marketing practices more widely.

More generally, this is a story of innovation processes from idea to market, and how actor-networks are sought to be built and stabilised around new consumer products, thus contributing to filling the gap in the literature on in-depth and real-time studies of innovation processes in industrial settings.

Combining fish, fermentation and proteins

The idea of fermenting fish

Professor Erik Slinde had the idea of testing fermentation, making 'salami' out of fish, when thinking about industrial opportunities of fish, and, in 2000, he brought it up with some of his students. He wanted to contribute to developing 'value-added' products from Norway's rich source of seafood raw materials, thereby opposing the prevalent view in the fish industry that industrialisation of fish processing in Norway is almost impossible in competition with low-wage countries both in Europe and Asia. The idea of combining fish with fermentation technology from meat evolved in conjunction with a group of researchers from the Institute of Marine Research, Tine R&D, the Norwegian Food Research Institute (Matforsk) and the Norwegian University of Life Sciences. Their competencies within biological and food sciences, and their curiosity towards exploring new ways of developing industrial food production, made them very interested in the novel idea of applying 'salami technology' to fish. The potential of utilising milk proteins to stabilise the product strengthened its relevance for the Tine researchers, and helped legitimise such activities within the realm of a dairy company. In this initial phase, the Research Council of Norway found this to be a promising exploration of blue-green innovation, and allocated significant funding to a project that was conducted from 2001 to 2005, the Neptun project, in order to research the use of milk proteins for stabilisation of fish products.

Tine R&D had already been working on milk proteins for a while, and this opportunity to explore a brand new application and hence understand more about how proteins work, was indeed attractive to their research community. The surplus of whey[2] from the production of cheese appealed to the 'product optimisation' logic of Tine as an industrial actor. Opportunities for utilising – that is, creating economic value from – more of the raw material was encouraged, and the use of whey was limited. In May 2001, funding for a large project was granted by the Research Council of Norway. The Neptun project included funding a PhD on the topic, and was a general investigation of the application possibilities for proteins, for which both the Norwegian Food Research Institute and Slinde were project participants. The emerging corporate biomarine innovation strategy at Tine, and the emerging blue-green discourse within national science politics, provided support for working on biomarine issues within an agricultural research community. I will go somewhat further into this phase of idea development and the Neptun project in Chapter 4. Here, I will provide an overview of how the project evolved further into a product development project, Umi No Kami, and then into attempts at commercialisation in the concept of the 'Salma' brand (see table 1 for the chronology of the case study).

Starting product development

During these initial experiments with creating a salami-like product of fish, Slinde filed a patent application. In this way, the idea 'took on reality', or became an object that could be sold to the food industry. As a consequence, this combination of fermentation and proteins was connected even stronger to Tine's biomarine strategy, thereby reinforcing and co-creating it when Tine's corporate management chose to purchase Slinde's patent application, taking full responsibility to develop and commercialise products from this novel technology.

Tine's initial rationale for considering doing something on the commercial side of this idea was the utilisation of milk proteins for stabilising the product, not the biomarine innovation strategy. Further, it was not until this strategy was in place and implemented during 2001, that it became relevant enough to buy the patent application. In November/December 2001 a deal was negotiated between Slinde/ ForInnova (University of Bergen's technology transfer office/TTO) and Tine. Hanne Refsholt, later CEO at Tine, recalled the event as part of a more systematic search for projects to include the recently developed

Table 1   A chronology of the case

1993–2003	Development of pre-rigor technology (Bremnes Seashore/ University of Life Sciences)
1999–2000	'Tine 2005' corporate strategy is developed, including biomarine innovation strategy
Autumn 2000	Professor Slinde tests the idea of fermenting fish with 'salami technology'
Autumn 2000	Slinde and Nordvi (Tine R&D) start informal cooperation on fermentation
11/2000	Slinde files a patent application for 'fish salami'
05/2001–02/2005	The Neptun project receives funding from the Research Council of Norway
12/2001	Tine buys Slinde's patent application, establishing the Umi No Kami project
Autumn 2002	The lab is moved from the university campus to Tine R&D; mould problems arise
2002	International market exploration tour
01/2003	First successful production of 'fish salami' without mould at Tine R&D's lab
06/2003	Market surveys in Norway
2002–2003	Tine hires researchers from University of Life Sciences
Winter 2003	Tests are conducted with 100% pre-rigor salmon in the recipe
12/2003	Cooperation with Bremnes Seashore
01/2004	Tine Biomarin and Tine Ingredients merge
Winter 2004	The Umi No Kami project moved to the line organisation
05/2004	Partnership forms between Tine and Bremnes Seashore: establishing Salmon Brands
06/2004	The Salma concept and brand is established
06/2004	Production at Bremnes is scaled up
08/2004–10/2004	Second international marketing tour
Winter 2005	Test sales of Salma Cured at KaDeWe in Berlin
03/2005	Agreement for test sales of Salma Cured in German retail
11/2004–08/2005	Negotiations with and adaptation to MRC (failed)
Spring 2005	Expanding the Salma concept with 'Salma Fresh' loins
Summer 2005	Receives design award for the Salma concept
Autumn 2005	Retail distribution of Salma Fresh in Norway is rolled out (first supermarket in September)
05/2006	Salma Fresh out in 20 supermarkets in/around Oslo
Summer 2006	Second attempt at selling Salma Cured in Germany (too low sales, withdrawn)
Autumn 2008	Salma Fresh out in 250 supermarkets and 40 restaurants in Norway
Autumn 2008	Salma Fresh out in 50 supermarkets in Germany

blue-green innovation strategy, but one in which they had to give up on the 'open evaluation' of incoming proposals:

> We got around 150–170 proposals, and then we had to make a system for handling the requests. And then we understood that sitting here, responding to proposals was not a very wise way to work. We had decided not to do fish farming, we would rather build on our own strengths, hunt synergies, and this became the guiding principle. When looking for synergies, some things just came along, because we had been involved for a long time in this project on fermenting salmon and binding fatty acids, long before any company was established. (Hanne Refsholt, CEO at Tine)

The approach of systematic scanning became too difficult. How could they know what was behind the proposals, with regard to competence and capacity? In part they ended up evaluating the potential of some of their ongoing activities, and this 'fish salami' idea came up as one of them. Hence, we can say that existing activities and competence, as well as their recent history, was influential with regard to what biomarine investments they made.

A product development project, Umi No Kami,[3] was started at the beginning of 2002, aiming for the commercialisation of the knowledge and technology developed in the Neptun project. In January 2003, they finally succeeded, at least to some extent, with developing and stabilising the technology. However, they did not succeed with finding and deciding on a final (stable) shape of the product until testing the unique salmon raw material from Bremnes Seashore in the recipe during the spring of 2003. On the market side, they had less success in choosing a particular direction, and with maintaining support from top management. Moreover, rigid preconditions for the project by top management were increasingly experienced as being a barrier to progress and finding direction.

Market research and technological development

Technically and visually the product was like salami, but it had been previously impossible to make salami out of fresh fish. New technology that combined techniques and materials from the biomarine and the agricultural domains had made it possible to solve two major problems in stabilising fish for such purposes. First, fatty acids in fish are too liquid, failing to provide the preferred stable consistency one would expect from such a product. Second, fish fats normally turn harsh in just a few days. Thus, there had been two problems of stabilisation in solving the technological basis for this strange new product. Still, there was more to it, including a number of stabilisation issues to be handled on its shape, taste, quality and conceptual identity. The transformation of a patent into something that could actually be industrially produced, and that was edible, involved agricultural bacteriology and chemistry, biomarine technology and biology, marketing expertise and management coordination.

When the Umi No Kami project was formalised, it received fresh resources for working on two interlinked processes. First, there was the process of continued and strengthened technological exploration and development, exploring the technological opportunities and developing the feasibility of the technology and recipe. Second, there was the exploring and sketching of a product concept by seeking knowledge of consumers and their reactions to the product. However, when taking ownership of the invention, further innovation work was moved in-house, breaking off collaboration with expertise on micro-biological technologies, due to lack of 'trust in doing these experiments at the Norwegian Food Research Institute' (Per Magnus Mæhle, Tine), or a desire to protect their knowledge from competitors. This had consequences, both in increasing the development time of the product, and creating uncertainty in moments of determining its further direction. On the industrial side, the search for potential partners started early. Even so, their main conclusion emphasised the following: 'the group considers the partner search to be less important than the product development. Without a physical product, negotiations with potential partners become shadow boxing' (from project meeting summary). Yet, with regard to product development, then, where do you start, when both the product and its market are unknown from the outset? It was a lot more difficult to mobilise the marketing department for shaping a product that was far from their existing set of food categories. Still, they emphasised several benefits of doing everything in-house at Tine. First, there was increased flexibility from being able to produce whenever one wanted, independent of available capacity in the Norwegian Food Research Institute labs. Second, economic arguments were mobilised. Third, it was argued that it improved the protection of new knowledge, in comparison to the more open environment at the Norwegian Food Research Institute.

In their earliest commercial analyses, the team emphasised the inherent paradox of making a mixed and cured product of fine raw materials that would potentially not be perceived as being of 'premium quality' by consumers. Still, it was described as a product fitting within Asian and European food trends, to be used up-market as pre-course, snacks or sandwich filling. In the first marketing plan, 'curiosity and health' were mentioned as the anticipated main triggers for customers, linking the product's content of omega 3 to the growing health trends within food. In the business plan from the same period, a success scenario was described: 'It was a real innovation... We knew that if we could manage the idea right, we would be able to launch a unique product concept. Not just in Norway, but also internationally' (from 'Business plan').

A mass market was projected, imagining how the 'salami' would become an everyday product 'on all breakfast tables and in all sandwich outlets'. Integration of development and marketing was thus depicted as both necessary – for understanding and meeting the user demands, and difficult – for the lack of methods for involving users before having produced 'something concrete' to represent the project. This dilemma led to a rather defensive strategy related to industrial customers (e.g., retail). Nevertheless, unlike other sites of research, an industrial organisation necessitates a business focus, and so the project group went out travelling, to Italy, Belgium, Japan and South Korea, to learn about their 'food cultures' and test their responses to very early versions of the product.

After a few years of the expensive and time-consuming course of events named the 'biomarine innovation strategy', Tine's owners, board of directors and, finally, management started to express increasing impatience with the economic side of the different projects. It also represented a considerable push for speeding up the commercialisation of the Umi No Kami project. In this setting, new opportunities coming from newly hired researchers from the University of Life Sciences and their relations to a fish farm, Bremnes Seashore, were timely, both for the management of Tine Biomarin and for Bremnes Seashore.

Including pre-rigor salmon

The interest in improving and developing processing technologies of salmon brought the fish farm, Bremnes Seashore and a group of aquacultural researchers at the University of Life Sciences together early in the 1990s. Throughout several years of collaboration, novel technologies for processing had been patented that enabled so-called 'pre-rigor' processing to be possible with great results documented on quality. These were new slaughtering and processing technologies of farmed salmon, reducing stress levels of the fish, and enabling pre-rigor[4] processing of the fish to an extent that no competitors could achieve. In addition to the advantage of time, and getting fresher fish out to the customers, the raw material proved to have some new and very interesting characteristics regarding colour, texture and gaping.

In the Umi No Kami project, Tine quested for high-quality raw materials; in other words, seeking to control the practice of suppliers. Gradually, and in learning what specific knowledge they lacked, they supplemented the team in 2002–3 by hiring aquaculture scientists and product developers from a research group at the University of Life Sciences. First, they hired Lars Petter Swensen for the Umi No Kami project, to help them develop methods for sorting salmon based on their fat content (high variation on salmon), which again would make it easier to control the fermentation process in the fish salami. His supervisor at the university, Per Olav Skjervold, and two other senior colleagues (Svein Olaf Fjæra and Odd Ivar Lekang), were brought in for coordinating all the biomarine research activities through a programme called 'fiskekraft'.[5] Eventually, they also hired Even Manseth, who was at the final stage of his PhD investigating the characteristics of some components in the blood of salmon. This new group of people created new dynamics in Tine's biomarine activities, and with their close relations to Bremnes Seashore this group had easy access to pre-rigor raw materials of superior quality. Subsequently, Swensen and Skjervold informally started testing Bremnes' pre-rigor salmon in the Umi No Kami –recipe, with good results. In Swensen's view, this change from using frozen fish (white and red) to fresh salmon was a seminal breakthrough of the project.

Meanwhile, Gunnar Hovland had been hired as director for Tine Biomarin. Further, when Hovland and others at Tine got to know and taste the superb quality of Bremnes Seashore's pre-rigor salmon, they became very interested. By teaming up such a partner on the supply and production side, they managed to stabilise a number of technical issues, but still had the problem of moving the innovation closer to commercialisation.

Shift from Umi No Kami to Salma

After the research group from the University of Life Sciences were hired, they contributed with new resources on the fish side that had an important influence on the further direction of the project. More than with scientific knowledge, they provided an overview of the fish sector and relationships to several competent actors within the industry. Later, then the pressure for economic results gained force, the new impulses from this group and their relations to the fish sector took part in moving Umi No Kami towards its commercial identity as 'Salma'. An alliance with Bremnes Seashore was established, and production facilities put up at their place, just 20 metres from the sea.

Hovland had got the job of cleaning up and pushing the biomarine activities in Tine towards commercialisation, and, in January 2004, Tine Biomarin and Tine Ingredients were merged, with Hovland as managing director, in order to gain synergies from Tine Ingredients' commercial capacities and network to business customers. The impatience with having a lack of conceptual, and hence commercial, progress in the Umi No Kami project led to moving the project from Tine R&D to 'the line', radically downsizing the project organisation, hiring a commercialisation manager, Øyvind Kiland, from Tine's marketing department, and getting approval for totally new preconditions from corporate management. The unique freshness of the raw material was highlighted in the concept called 'Salma', and in the packaging design. This was where Tine and the Norwegian fish industry were to become unbeatable, as this quality level could only be produced close to the sea: 'We are depending on fresh raw material, so it can only happen in Norway. It is unique, and this is where the future of Norwegian fish industry is found, in hyper-fresh processing' (Øyvind Kiland, Tine Ingredients).

However, the differences between aquaculture and agriculture related to 'culture', competence and economic/market systems were frequently mentioned as major challenges by the project participants. Moreover, the interests of the local family business (Bremnes Seashore) and the large corporation (Tine) were different. Nevertheless, they succeeded in negotiating an agreement of a strategic and long-term alliance, and the work of establishing production practice at the farm could start.

Scaling up production

The new technology had to be made more robust in order to endure the transfer from small-scale laboratory production to large-scale industrial production. The transfer and scaling up of production from the dairy lab to the large-scale facilities at the fish farm was done during the summer of 2004 (see Figure 1). Yet then, a few weeks later, problems occurred. Suddenly, the whole batch was attacked by mould and had to be discarded; overnight the harmony was gone. A white layer of mould had invaded every salami in the drying facility. This was also challenging for the marketing people, who had already started presenting the product for various international customers. Through training and reinforcement of standards and routines, the problem disappeared and

Figure 1   Salma Cured

did not reappear in later productions. However, this mould incident should not have been too surprising for the Tine people, as their own people had previously spent six frustrating months fighting mould in the their own laboratory. Unfortunately, it had all been forgotten in the black-boxed early history of Umi No Kami.

The next technical task was to improve the practice of several more aspects of micro-biology. In contrast to the fish industry standard of a maximum bacteria count of 10,000 per gram, they managed to decrease the bacterial concentration to below 1,000 bacteria per gram after the salmon was ready processed and packaged. Hence, they could guarantee the quality of the product with a minimum shelf life of ten days, producing economic and organisational advantages related to logistics and consumer experience. This was achieved through systematic collaboration over several months, with the production personnel, the cleaning personnel and the management. The commercialisation manager also related how he, on a couple of occasions, had to rush out to stores and withdraw products – for instance when products had started showing black spots, resulting from the slaughtering of too stressed fish (which led to blood shooting out to the muscles – something that did not appear until a few days later, when the blood had coagulated). In his view, no variation on quality or on the visual presentation of the product could slip through the control when building a high-quality brand.

Market testing: presenting the product

Three years after the initial idea the time had finally come for a market test. This meant intensive work on the naming, categorisation

Figures 2 and 3   The first Salma package

and designing of the product, which was based partly on projections of 'ideal' users, and partly on direct interaction with potential users. The change to pre-rigor salmon and a high-quality production strategy enabled (and required) a high-end concept. While designers worked on shaping the visual *representation* of Salma (see Figures 2–3), elite chefs worked on its *presentation* (see Figures 4–9).

For an object like Salma, its audience would easily become uncertain about what this was meant to be. Thus, figures showing Salma on display among some of the trendy 'hipster' types of food were produced – for example, sushi and seafood pizza. Later, some of these chefs would also appear at various places where Salma needed exclusive introduction to new actors, such as food fairs and business meetings.

Next, Salma started to assemble a set of associations, making it presentable as a viable concept for conscious consumers willing to pay to try a healthy alternative.

Market testing: selling Salma Cured

An international marketing tour was done with Salma in autumn 2004 and winter 2005. Existing business relations, food fairs and new contacts were visited in US, France, Singapore, Brussels, Moscow, etc.

Figures 4–9   Use situations for Salma Cured

Feedback from and interaction with different actors in these locations came to have a great impact, both on the salmon salami, and on the future commercial strategy of Salmon Brands, with the adaptation of the sausage in Asian pizza restaurants, German retailers and in Norway, which involved a reworking of the entire innovation for the 'home market'. While visiting Hong Kong, the team met representatives for a multinational restaurant company. This company was seen as the 'ulti-mate customer' for Salma at this stage, representing everything they hoped for: restaurants (easy logistics), worldwide distribution and asso-ciation with acknowledged brands. However, this customer also had some labour-intensive demands and, to be able to properly answer the question of feasibility for warm food, Salma had to be taken back to the laboratory. From (finally) being stable both in shape and production, its identity was again in question, or opened up. After altering some of the steps in processing, the results were positive. Unfortunately, in the meantime, the customer had, for unknown reasons, lost interest, and the attempt to mobilise the desired customer had brought about a great deal of work and a failure.

After Tine's agent for the distribution of cheese in Germany, Detlef Martens, expressed interest in Salma at the food fair in Paris, plans for distribution to retail chains in Germany started emerging. Very early, it was tested at the prestigious hypermarket, KaDeWe, in Berlin (winter 2005). Although it did not become an immediate big-selling hit, it received positive feedback from consumers, especially when it was demonstrated in the store. Second, it was sold on the ferry between Oslo and Kiel, where it sold steadily to German tourists wanting to bring something Norwegian back home. In March 2005, an agreement was signed between Martens and Tine for test sales of Salma Cured in German supermarkets. However, Martens was uncertain about the suitable categorisation of the product, related to the shopping practices of consumers. While emphasising its similarity with meat products, he still chose to locate the product together with smoked salmon and other cured seafood products. A number of purchasing managers were convinced and willing to give it a try, so Salma was launched for test sales in 90 German 'hypermarkets'. However, the sales of the 'Lax Salami' did not go particularly well, leading to it being put on hold.

Market testing: selling fresh

Salma had still not found its final shape, and no closure could be achieved before making some real and voluminous sales. Strategic considerations of brand development and positioning and decades of experience with the food industry could not settle Salma's identity. Its fate was fully in the hands of the customers (industrial actors) and their customers (consumers). Hence, it was easier to go back to the laboratory and the marketing department to develop new versions of the product, particularly products that came closer to already existing products in the market. At this point, a couple of ideas that had been considered for a while gained strength. Under pressure for economic results, the idea of marketing the fresh salmon loins instead of curing them, and the idea of working with Tine's established market relations in their domestic market got full support. As opposed to the salami version, the marketing of 'Salma Fresh' was launched in Norway, thereby starting in a familiar setting, where Tine already had relations, recognition and a strong market position with several other brands and products (dairy

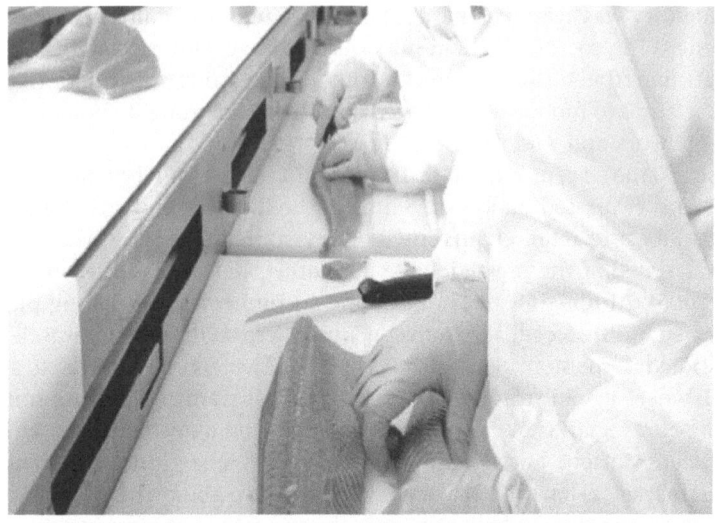

Figures 10 and 11   Salma Fresh

and ready-meal products). Neatly cut loins without skin and bones were packaged in transparent foil and with the same minimalist design concept (see Figures 10–11).

The strategy was to start with the best stores; Jakob's, a 'gourmet supermarket', immediately caught interest, and agreed on a test campaign in September 2005. The consumer response was very good. Because the supermarket was associated with a large retail chain, NorgesGruppen,[6] a long-time customer of Tine's dairy products, access to nationwide distribution opened up. Distribution of Salma Fresh was gradually rolled out in Norwegian and German supermarkets, as well as a number of high-quality restaurants, from 2006 onwards, making a success story of Salma.

Summing up the case study

These are the three 'programmes' related to the innovation process(es) to be described and analysed in the next chapters. First, there is the management of Tine's biomarine innovation strategy, and subsequent investments. From the start, this constellation of people from corporate management and Tine R&D was also responsible for formalising and mobilising the Umi No Kami project through buying the patent application from Erik Slinde and ForInnova. Later, under increasing pressure from their owners for proof that the blue-green strategy was viable, and then being presented the raw material from Bremnes Seashore, the new ideas about how to make profits from the biomarine area gained ground. Second, I describe and analyse agricultural and micro-biological researchers in the research network between Tine R&D, the Norwegian Food Research Institute and the University of Life Sciences, who sought to understand more of the workings and combinations of proteins, bacteria cultures and fatty acids across the different kinds of raw materials (meat and fish), as well as realise a fermented fish sausage. And third, I describe a small fish farm on the west coast (Bremnes Seashore), which committed to greater technical innovation than average, collaborated with another scientific community from the University of Life Sciences (aquaculture oriented) and struggled to create economic value from their technical improvements. Still, these programmes and projects did not evolve in isolation, and I will now outline some aspects of the settings in which this happened. In particular, I focus on the role of agriculture in the emergence of 'blue-green' innovation, and Tine's development of a biomarine innovation strategy.

Blue-green innovation and Tine's strategy

Agricultural R&D[7] and the appeal of fish

Tine has a long history of industrialising milk and related products. They have the largest R&D department in the Norwegian food industry, a national distribution system and one of the strongest market brands recognised by Norwegian consumers. Yet it is also a cooperative, owned by 18,000 farmers in a totally dominant position, with price regulation responsibilities and a role in rural politics. Organisational boundaries between Tine, the Norwegian Food Research Institute (Matforsk) and the Norwegian University of Life Sciences have been blurred, as there have been many collaborative projects, recruitment of people and cross-utilisation of facilities and other resources between them.[8] There are both historical and economic reasons for this. Historically, in 1924 the law on 'quality control' was approved, and the 'cheese-making' regulations arrived in 1925, thus driving dairy practices towards the standardisation of production processes. In relation to this, the Dairy

Cooperative and the Ministry of Agriculture and Food developed the four-leaved clover as a quality mark for products approved according to this standard. Later (1972), the 'Tine' (a traditional wooden box) was chosen instead (Espeli et al., 2006: 82); gradually the symbol for quality in terms of industrial standardisation and control became *the* identity of the cooperative Tine. The Dairy Laboratory, established in 1934, became responsible for control, and the quality control of dairy products was increasingly based in laboratory testing. In 1940, these test results influenced the price of raw materials (raw milk), and the Dairies' Producer Services got the role of adviser on quality improvement both in local dairies and in individual farms. This work was intensified during the Second World War, and eventually all cheese produced in Norway was controlled and priced according to quality (ibid.: 84). At the Agricultural University College (now the University of Life Sciences), the Dairy Institute had an experimentation dairy ready for use in 1953. Here, they set out to understand the influence of cheese-making technology on a more unified quality of cheese, and in addition a few new types of cheese were developed – for example, the famous 'Jarlsberg', an emmenthaler-like cheese based on gouda technology. The bacteria culture for Jarlsberg was still produced there until 1991, when Tine R&D's unit at Voll took over, and where its 'filter fermentor' technology later came to be used in supplying new production facilities in the US and Ireland with the precious bacteria culture. The Dairy Institute handed over the name and bacteria culture to the Dairy Cooperative without any form of economic compensation in case of prospective market success (ibid.: 88). However, the trademark, the Tine symbol, was not registered before 1972. From 1968, there was more systematic collaboration with the Dairy Institute, while, during the 1970s, R&D went through a phase of reorganisation and centralisation in the Dairy Cooperative. The division of labour between the Agricultural University College/the Dairy Institute and the Dairy Cooperative was discussed, with the cooperative at that time having developed 'well educated staff of more than twenty dairy engineers and others with higher education, and in addition dairy technicians, laboratory assistants and secretaries' (ibid.: 204). Instead of co-locating R&D at the university (at Ås, near Oslo), the Dairy Cooperative's R&D facilities were set up at larger dairies, to secure practical relevance and low barriers between lab and large-scale production. The basis for product development was quality improvements on already existing products, resulting in a quality of traditional dairy products of a 'relatively high level'. Espeli et al. further report that 'in practice it is difficult to divide product development and quality development', and that this focus, in combination with the Dairy Cooperative's market coverage, has not led to much development of new products (Espeli et al., 2006: 203). A new R&D centre was built in Voll in 1979, with 40 employees. Nonetheless, quality was still the single most important issue, with product development only accounting for 10–20 per cent of the activities at the centre (ibid.: 205). As late as the 1990s, the focus of R&D was still on quality, which entailed the standardisation both of quality parameters and the product portfolio, and local variants of the products were forced to adapt to the standards (ibid.: 209). Production was the only focus for R&D, and by erasing the freedom of local actors to maintain tradition and meet local demand, R&D was taken away from one (local) type of market orientation, while failing to connect to other types of market orientation (regional, national, international, etc.). It was all about solving production-related challenges. Still, the external and, to some extent, market-related pressure during this period came from nutritional expertise and regulation bodies. The pressure for less fat content in the products and new products with health benefits was the strongest (external) driver for product development within the Dairy Cooperative (ibid.: 209).

Uncertainty and preparation for potential international competition on agricultural food products increasingly preoccupied the industry from 1990 onwards. In 1992, the Dairy Cooperative launched the 'TINE' brand, together with restructuration of its management. A new era of developing market orientation followed from the corporate management – for example, through segmentation towards more specified groups of consumers (e.g., the Litago concept for children launched in 1993). Still, the continuous decrease in milk consumption throughout the 1990s assured continuous pressure for production and raw material orientation in product development. The aim clearly was, with some success, to stem the loss in milk consumption (ibid.: 226).

Thus, a rude summary of the development of R&D resources at Tine is that they have evolved from dairy technicians controlling and advising about hygiene and product quality, through the academisation of these functions, into a full-blown R&D centre with state-of-the-art knowledge and technology within niches of microbiology, life sciences, production technology and sensorics. Until around 2000, however, the drivers of innovation have been quality control, production optimisation (exploiting more of the raw material, e.g., 'kesam' from the surplus of whey, a by-product of cheese production) and production capacity (utilising free capacity of facilities, e.g., idle milk facilities for fruit juices). Even though R&D had occasionally been involved in various side projects, the shift towards taking a more active role in researching and developing new technologies and products came with the 'TINE 2005' strategy in 1999/2000, in which 'new business areas' were put forth as part of the main strategy, with fish as one of the identified new areas for strategic innovation.

In addition to these historical/technical, economic and managerial factors, links to science professions have obviously influenced R&D at Tine; in particular, microbiology, production technology, sensorics and nutrition/health. Entering into a Tine lab immediately gives the impression of something between a natural sciences lab, a small-scale production facility and a professional kitchen. There are people with white collars and sterile caps and shoe protection looking into high-tech scientific instruments, steel pipes going in all directions under the roof, and various smaller and larger machines everywhere: a calm and controlled atmosphere without the stressful tempo of a produc-tion plant. This picture was further strengthened by close relations to the Norwegian Food Research Institute (Matforsk) and the University of Life Sciences. Agricultural R&D has also been strongly connected to science and rural politics – for instance, when the political and medical debate on nutritional practices from 1963–83 led to the development and eventually introduction of fat-reduced milk to consumers (Espeli et al., 2006).

Tine saw how decades of knowledge production on agricultural food quality (nutrition, hygiene, distribution and marketing) could possibly become a valuable development of fish processing. In addition, during the last decade or so, Tine developed their technical and professional knowledge related to product development in a much broader sense, from micro-biology (fermentation, bacteriology, fatty acids, proteins, water activity, sensorics, etc.) to conceptual development of products and packages together with designers and marketers, and their own cold-chain distribution channels. Fish had already been on the agenda a few times, either in relation to product development (e.g., ready meals for Fjordland), or to more basic research on the characteristics and application of various ingredients (marine omega 3 for dairy products, milk proteins for various fish products, etc.), and now they decided to make this a strategic priority.

Blue-green innovation

In this interaction between corporate strategy and R&D activities, a portfolio of biomarine projects emerged within the Tine group. During the same time period, several interrelated issues gained leverage in the public debate and among business and R&D actors. First, there was a governmental policy debate on 'blue-green' innovation, discussing how to restructure the R&D sectors related to agro-food and biomarine food in order to gain synergies from cross-sectoral developments. This resulted in a governmental white paper in 2004 called 'The Blue-green Food Alliance: Joint Efforts and New Structure'. This discourse was drawn upon by the different actors involved in planning and realising biomarine innovation in this case study. Second, from the 1970s onwards, a couple of research groups in Norway had taken leading roles in the successful domestication of salmon (and, gradually, a number of other species). One of the two leading Norwegian research groups on the domestication and development of the breeding of salmon was based at the University of Life Sciences,[9] a constellation of scientists from the breeding of cattle and pigs, aquaculture and experiment stations by the west coast. Professor Harald Skjervold is mentioned as a leading figure on combining and translating the breeding system and knowledge from agriculture to salmon breeding in the early 1970s. His son, Per Olav Skjervold has, since then, together with a group of researchers from the same network, developed processing technologies for pre-rigor salmon that play a central role in the case study presented here. Akvaforsk (Institute for Aquaculture Research), in Ås, is still a leading institution in its field. Farming fish instead of catching wild fish means a completely different regime in terms of control of raw materials – both regarding availability/volumes and qualities (Aslesen et al., 2002). Thus, the biomarine area has opened up for industrial production and marketing on a totally different level. The market system, however, had still not gone through a similar shift – thus, for the most part, refusing to reward 'value-adding' activities, such as product development, branding, etc. Third, also stemming from the optimisation logic of industrial production, an increasing number of scientists and entrepreneurs were focused on the search for 'gold' in waste (by-products) from fisheries. Around 50 per cent of the fish were normally considered waste, only 14 per cent of the fish were processed and sold as fillets and no more than 2 per cent of Norwegian fish were processed in Norway (according to Tine Biomarin). In other words, the potential could be huge both for further processing of fish, product development and for creating value from by-products. Products for food, as well as cosmetics and medicine have been developed and commercialised from various parts of waste from fisheries, shrimp factories, etc. These broader trends and discourses played a supporting role for the development of a biomarine innovation strategy and business development unit within Tine. The agricultural and biomarine sectors are more different than initially might be thought, in terms of knowledge and technical regimes, market organisation and hence in innovation power. These differences have historical, cultural and technical roots, from 'catch'- and 'cultivation'-based paths developed through centuries of human activity. At the time of writing this book, mainly generic products were sold in the Norwegian aquaculture sector, such as whole, unprocessed fish, while on the agricultural side, industrial processing, or 'value-adding', of raw materials was more the rule than the exception.

The Reseach Council of Norway is the main actor for channelling government funding of research in Norway. In a recent report on a number of the blue-green research programmes, both the science-political drive for integrating blue and green research and innovation, as well as ambiguity regarding the blue-green potential from practitioners was presented. The managing director of the Norwegian Seafood Federation (FHL), Geir Andreassen, evaluated his experiences with blue-green innovation:

> the value chain perspective (fish/agriculture) has possibly been a restraint to the development instead of promoting it. The conditions, opportunities and challenges in the agricultural and marine sectors are fundamentally different and may result in a vague and fragmented research focus.... We have an enormous raw material base in wild species and breeding. What remains is creating industrial activity to increase the value creation from these raw material resources. At the same time, we have to develop new knowledge on how we can exploit the same raw material base in relation to bi-products and ingredients in other industries. (Geir Andreassen, FHL, in RCN Research Report, 2006)

Although his concerns for the marine sectors were remarkably similar to Tine's arguments for going into the fish industry, Andreassen had little belief in any synergies emerging from collaboration happening across the agro-marine boundaries – as they are 'fundamentally different'. Noting that this was expressed in a report on the Research Council of Norway's (RCN) portfolio of blue-green projects, this has to be read as a rather strong criticism of the idea of blue-green innovation. From this report, we can see that most of the Research Council's blue-green programmes during this period were in reality either blue or green, but the two programmes mentioned by Andreassen sought to integrate blue and green, and were criticised for this. Andreassen's scepticism might also have something to do with not wanting to lose ground to agricultural actors, or perhaps with lack of knowledge of what 'the other' (agriculture) represents in terms of complementary knowledge and resources? The coordinators of these programmes in the research council, on the other hand, were more optimistic, and wanted to see more integration: 'We want to get even further in finding synergies between agriculture and fisheries/aquaculture. We succeeded with this, especially in the "food program", but we know that it is possible to accomplish more' (Danielsen and Horn, in RCN Research Report, 2006).

Clearly, there were conflicting views on the results, and perhaps even the aims, of these programmes. According to the programme chair of the 'market and society' programme, Abraham Hallenstvedt, professional boundaries were part of the problem for achieving more interaction between blue and green:

> if we increasingly could get researchers with experience from agriculture and forestry to work together with researches from the fish sector and formulate common research questions.... Researchers thrive in their own sectors, and I don't think they should be allowed to continue like that without intervention from the outside. (Abraham Hallenstvedt, in RCN Research Report, 2006)

While the whole report was designed and structured *as if* there were blue-green synergies in the programmes, thereby presenting blue and green projects side by side, the only concrete example in the report of something actually blue-green is the 'super cooling' project in the 'food programme', where Sintef, Gilde (the Norwegian Meat Cooperative) and the Norwegian Seafood Federation together developed new cooling technology and then also transferred it to fish, enabling the cooled storage of products at –1.1 degrees Celsius, increasing durability and quality to the consumer. All in all, the report indicates that there was still a long way to go before the blue and the green could be said to be integrated in any sense. To co-locate and coordinate, does not automatically mean to integrate.

Tine's efforts to industrialise fish

For Tine, there were at least two central and relatively new problems arising in the next years to come. First, they were likely to face more competition on their core business from large multinational companies with a lot more experience on how to survive in an 'open' economy. As EU and World Trade Organisation (WTO) negotiations evolve, it is becoming clearer that their privileged national market position will not last forever. According to the director of research at Tine R&D, Johanne Brendehaug, Tine was well aware of the potential competition challenge, exaggerated by a different cost structure of milk farming and production than in neighbour countries: 'It costs 3 NOK to produce one litre of milk here, while it costs around 2–2,5 NOK per litre in Sweden. This is a threat to the food industry. Therefore, we have to compensate with knowledge development' (Johanne Brendehaug, Tine R&D).

Olsen and Espelien (2003) have argued that the lack of investments in infrastructure and production facilities in the Norwegian farming system have contributed significantly to these differences in cost structure. Thus far Tine has been able to compensate for the significant national decrease of milk consumption[10] for the last 15–20 years with increasing levels of product development, launching numerous new products (desserts, yoghurts, beverages, ready meals, etc.) to the markets each year, so that their turnover and profits have gradually increased during the same period. This suggests that Tine has managed to transform and develop its R&D and marketing competence towards a more 'innovative' profile.

Joint knowledge development across the sectoral borders of agri- and aquaculture has during the last three decades achieved considerable success in a few areas. The strongest example is perhaps how knowledge about breeding, feeding and the health of cattle and pigs has been utilised to develop the present world-class quality and cost efficiency of Norwegian salmon farming, through joint research based at the Norwegian University of Life Sciences. However, this bridging of the two sectors has been limited, and Tine has, through their blue-green innovation strategy, tried to take it several steps further, into the industrialisation of food and ingredients production.

Two significant changes at Tine are of importance for our understanding of this case study. First, the implementation of a corporate organisational structure was decided in 2002, representing a relatively dramatic shift in mindset from the long tradition of regional cooperative organisation. Although not breaking off from the original idea of cooperatives, this shift signalled centralisation and a much stronger adjustment to the corporate world. Gilde (the Norwegian Meat Cooperative) had taken the responsibility of moving first. After a few years of systematic activities in informing and dialoguing with their owners (farmers), the traditional group of the regional meat cooperative was enrolled into a single corporation in 2002, and then Tine went through with the same decision shortly after. Second, due to both this corporatisation process and the strategy process of 1999/2000 leading to the 'TINE 2005'[11] strategy, Tine was no longer to be just a dairy company, but first and foremost a 'food corporation'. Its new statutes were formulated in more general economic and food industrial terms:

> TINE's objective is to run an effective, quality and market oriented business in the food industry on a cooperative basis. TINE shall on behalf of the owners strive for the best possible economic result from their milk production, and in addition take care of the owners' other common interests. TINE may also import and export products. TINE may through investments or in other ways take part in other business. (Excerpt from Tine's new statutes, 2002)

These formulations clearly opened their field of activity, both within and beyond their current business. In commenting upon this change, corporate director Per Magnus Mæhle framed it as an 'expansion' of their knowledge:

> This has to do with an expansion, and it became inscribed in the statutes when the corporate model was implemented. Now this is not to say that Tine's purpose is to produce milk, or process milk, but the primary thing is that Tine's purpose is to be a central actor within food production in Norway and the Nordic region. (Per Magnus Mæhle, Tine)

This change is significant, from being a Dairy Cooperative whose only scope was to process, distribute and market the raw material of the owning farmers. The chain of arguments making this possible can be read in the above statement: changing the name of the cooperative from 'Norwegian dairies' to 'TINE'[12] and from milk to 'food industry', including 'economic results' and 'other business'. Two years earlier (2000), the 'TINE 2005' strategy was implemented, in which the biomarine sector was chosen as one out of four new strategic 'innovation areas' to work on in the coming years. Espeli et al. also argues that this new field of innovation became possible partly due to the corporatisation of Tine:

> Establishing the corporation in 2002 implied an opportunity for the integration of businesses that formerly had been managed without national coordination from the central administration. The corporation also provided opportunities for engagements in totally new areas, like seafood. (Espeli et al., 2006: 312)

The new identity as 'food corporation', instead of dairy, opened up a number of potential business areas, such as fish, ready meals, bread products, sauces, desserts, soups, spreads and cold cuts, snacks, chocolate/sweets, pizza, baby food, brand extension, co-branding and new drinking products. Moreover, animal feed, patents, commercialisation of brands and licenses, and capital/property investments were considered (ibid.: 290). Potential synergies were outlined, related to competence, distribution system, supplier and customer networks, production facilities and branding. Seafood got a particular role (ibid.: 289). With a mandate from the corporate management, a group for developing the 'new business areas' strategy was established,[13] and the work of this group would come to set out many of the premises for discussing the traditional dairy path versus the new food path (ibid.: 292).

It is reasonable to say that Tine's product and technology competence over the last 70 years or so had mainly been based on quality development (i.e., hygiene, animal breeding, cold-chain supply and distribution), while from the turn of the millennium, product development received a more strategic focus. While working on all aspects of product development, from biotechnology to concept design, the utilisation of exactly this 'quality competence' has arguably turned out to be valuable. In order to develop the quality of processing of, for example, fish, enabling the production and branding of high-end products has been a powerful way to do product development.

Similar to the global restructuring of the food retail sector, the restructuration of the Norwegian food retail sector during the 1980s and early 90s resulted, in 1995, in four big actors totally dominating the market, and building their own integrated systems of business and distribution. There has thereafter been a continuous fight over distribution between these retail chains and the agro-cooperatives. Yet, on their cold and fresh products the agro-cooperatives did not yield an inch. Their cold-chain technology and distribution system was said to be an important strategic means for keeping control over the market (Espeli et al., 2006: 327). Moreover, while in the 1970s and 80s every new product from Tine would be warmly welcomed by retailers, the new powerful constellations of retailers saw things differently. Central coordination in the retail chains of product assortments increasingly put constraints on the potential for product development, especially after 2000, forcing Tine to involve their (industrial) customers much earlier in the product development process (ibid.: 329). According to Jan Ove Tryggestad (board member at Tine and several of its biomarine companies), these changes have created friction in the organisation. As an early example, the idea of utilising available technological capacity for producing fruit juices met resistance in the organisation, as people were afraid that it would take focus away from dairy/milk activities, and potentially cannibalise the milk market.

This understanding of Tine as a politicised organisation was strengthened by the findings of Huse and Schøning's (2005) study[14] of the Tine board of directors in the same time period; a collectively owned organisation, with board members representing different positions regarding both Tine's strategy and the politics of farming and food production. This political aspect of the organisation holds important relations to the larger networks in which Tine is embedded – to more than 18,000 Norwegian farmers, to the national and regional politics of food production, to the rural population and to national protection of the agricultural food market. Some of these relationships came to play a role in relation to the issue of 'blue-green' innovation.

As mentioned previously, there are significant historical, technological, economic and cultural differences between the agricultural (green) and biomarine (blue) sectors. The more interesting question, however, seems to be whether these differences mean that the sectors should be kept separate (no potential for synergies) or if a stronger relation would pay off both technologically and economically. Professor Erik Slinde (Norwegian Institute for Marine Research), inventor of the 'fish salami' in this case study, explicated the differences between the sectors in this way:

> Farmers know that the cow needs a calf in order to produce milk, and that the calf must become a heifer. Farmers think long-term. Most people complain about farmers, because they get a hell of a lot of subsidies, and why do they? They are incredibly clever negotiators. Fishermen, they live from the principle 'this year our God is kind, giving us lots of fish', and 'this year there are few fish, so we have to starve'. They think extremely short-term. And there is no hope for the traditional fish industry. (Erik Slinde, Institute of Marine Research)

In spite of its exaggeration, this statement represents some of the basic differences between fishery and dairy. A catch-based business, like traditional fisheries, has not allowed for long-term industrialised development of knowledge, technology or markets in the same manner as the agricultural sector. Further, while the fish sector is transforming towards cultivation instead of catch, industrialisation of processing, product development and marketing are still in need of development. This is what the corporate management at Tine identified as an opportunity.

Moreover, Tine has latterly had to revise and reverse these considerations of not covering the whole value chain, realising that issues related to farming perhaps represent some of the strongest potentials for synergy between the blue and the green:

> And then we said, in opposition to the milk area, we will only be where we think we can get added value, and where we think we are able to contribute as close as possible to the market. But we have acknowledged that this is impossible and wrong, because, similar to milk, it is as vulnerably related to quality control in the whole value chain. This means that we have to go in and take control, even in raw material supply and processing. (Per Magnus Mæhle, Tine)

Integrating backwards in the value chain has, especially in the case of this study, become a success factor. Although they have not yet gone all the way back to breeding fish (with the exception of some research projects on fish feed), they now seek to influence everything from when the fish is taken out of the water to its presentation in supermarket shelves and restaurant menus. This is not just a simple operation, as it demands both general and specific knowledge:

> You can't just know something about organic components in food products, you must also know the specificities of fish as raw material which are different from what we know from before. And you need to know about the market, sourcing-wise, how it is different from ours, and such issues. (Per Magnus Mæhle, Tine)

In other words, Tine had learned from its biomarine ventures that it needs to know more about the specificities of the new sector in order to succeed with using/applying its own expertise to that sector's raw materials, products and markets.

It turned out that the technologies and methodological practices of research and development could be transferred with some adjustments, but the specific ways that the new raw materials would behave, their 'characteristics', were different, and here Tine sought external expertise. While the simplification, or generalisation, of breaking materials down to their more fundamental (and thus similar) pieces broadened the scope of their knowledge and enabled the transition, this did not mean that they automatically would succeed. It could well be the case that their generalisable knowledge would not handle the particularities of the new area. The director of product development at Tine R&D, Ove

Johansen, combined the new strategic inclusion of 'side activities', like the biomarine projects, with their generalised R&D competence and their ultimate goals of maximising economic returns to Tine's owners (agricultural farmers):

> We are supposed to maintain the income of our owners; they should be able to make a living from the raw material they provide. But now this means first and foremost that they make money, no matter if the income stems from milk or fish or other things.

The dairy cooperative organisational model was interpreted quite freely here, suggesting that the main issue was not only about processing, distributing and selling their owners' milk, but rather about providing them with economic income. 'Income from milk' was translated to 'income from food'. Again, a generalisation of Tine's activities, competences and mandate took part in enabling the move to blue-green innovation. In addition, the fish industry was projected as being less competent and hence in need of Tine's expertise:

> The production processes of Maritex and Tine are quite similar, and we can make them better. The fish industry has no tradition for good hygiene, it is pretty rough. We can use omega 3 in Tine's innovation. There is a problem with fish oil becoming harsh and with durability, but our knowledge can be used for handling these challenges. (Ove Johansen, Tine R&D)

This view, of the fish industry as being 'pretty rough', with less emphasis and looser regulations on hygiene and other 'nutritional standards', is representative of most of the agricultural actors (both in business and science) I have been talking to. On the one hand, this is based on some evidence, or at least experience, while on the other, this serves as the basis for supporting and enforcing Tine's strategy of taking on a position in what they believed would become an industry more similar to the agro-industry than it was previously. Within an integrated food producer like Tine, the strategy could then be to become the first to develop and market combined products in their own product portfolio, thereby enabling the marketing of a whole 'package' of application knowledge, technology and actual ingredients to other industrial actors internationally.

At the outset, the blue-green investments were meant to be long-term, meaning they allotted three to seven years before the projects needed to produce economic value. At the same time, there was a continuous discussion of how patient one should be, what kind of resources should be made available and what intermediate goals to push forward. There were certainly not unified discussions in the board and management of Tine, nor in the daughter companies.

The biomarine projects

In going further into the development of a blue-green innovation strategy at Tine, the implementation of the strategy can be viewed as a broad trial and error exploration of a new field for the involved participants. Two main routes were tested, the financial and the industrial, of which the former had a short career. The financial position of the investment venturing company INAQ, formed by former consultants in KPMG, together with Tine, Gilde and some shipping actors, ended in failure: 'INAQ was an early case we took on, and it was more for financial reasons, plus to build a network in the sector and seek out a way, and it was not very successful, therefore it got closed down' (Hanne Refsholt, Tine).

The initiative came up in established relations within business consulting, and was boosted by the optimism of the emerging aquaculture 'revolution'; according to corporate director Per Magnus Mæhle, later it was undermined by the subsequent downturn in the industry:

> *Mæhle*: INAQ is to date the absolute most unsuccessful venture we have done.... It was started by a group coming out of KPMG, and their sea farming group, and we were among those joining in on investing in the company when it was established, together with some others. Gilde joined in, as well as a couple of ship-owners, and some other investors.
> *Researcher*: What didn't work out in the project?
> *Mæhle*: Well, it was stopped by the downturn in the breeding sector. Failed investments and over-investments plus a conflict where one of the owners on both sides of the table, when buying a company with no value.

Hence, instead of learning about the fish industry, or building alliances for further investments in the sector, what they learned was to stick with their core competence, and directly involve industrial production and marketing of food. Accordingly, the other main route was one of industrialisation. A set of very different projects was established, with the shared idea to build and exploit industrial synergies with Tine's existing business: knowledge, technology, distribution and marketing. The common coordinating organisation, Tine Biomarin, was established in autumn 2001:

> Someone who knows Tine well enough to be able to draw on relevant resources, directly related to the individual ventures. Later, we chose to close down Tine Biomarin, because we questioned the value creation in Tine Biomarin. We saw that it didn't manage to become anything more than a hub; it didn't reach a critical mass in terms of the marketing and sales organisation. (Hanne Refsholt, Tine)

In order to ensure access to the relevant in-house resources at Tine, and to facilitate interaction between the different projects, they were organised under a common organisation. However, later they saw that this organisation served this 'hub' function only, and lacked the ability for 'value creation' (commercially). Thus, when they chose to merge it with the supplier of ingredients to industrial customers, Tine Ingredients, it was important to maintain this 'networked practice'. Still, not all biomarine projects were located within Tine Biomarin, as some projects related to the distribution of fresh fish were organised closer to the daughter company, Fjordland, a producer of ready meals, margarines, etc. One example is the case of 'Marian Seafoods', a concept of fresh fish distribution to Norwegian retail:

> We first established it as a project together with Gilde, and worked a couple of months together with them, before we found out that we needed a partner from the fish side, and got Domstein and Fjord Seafoods partly involved on the ownership side. But it turned out that we didn't have similar enough strategic goals, we simply could not agree. (Per Magnus Mæhle, Tine)

Sectoral boundaries, here in the form of different interests and goals, made it difficult to collaborate in a strategic alliance with actors from the fish sector. After this breakdown, Tine and Gilde together established the company Marian Seafoods, with the concept of packaging fresh fillets of fish with packaging technology from the meat industry, and aiming for the large Norwegian retail chains. Later, the company was merged with Fjordland. However, long-term problems in getting a stable supply, both in volumes and quality, especially on white fish dependent on catch, created a great deal of trouble.

> One thing is to secure access to white fish and establish packaging facilities at Fosnavågen. We can reduce the degree of innovation, but

we can't compromise on process quality. They did not have the quality in their production that is called 'nutritional standard', and we
have had to improve this, and it has cost money and a change in
attitude. (Per Magnus Mæhle, Tine)

Quality, in the sense of hygiene and 'nutritional standards', seems to
be like a 'sacred cow' within the Tine system, a matter that has no compromise. Either they manage to get the quality they look for, or they
cannot do anything at all. After changing suppliers a couple of times,
a partnership with Norway Seafoods was established, one of the largest fishery companies in the country, presumably with the resources to
develop their supply according to Tine's standards. However, the market
side had problems, too. After some new attempts at launching Marian
products during 2005–6, together with a large campaign using expensive TV commercials without success, the concept was taken out of the
Fjordland portfolio and closed down.

The next large projects were within marine ingredients, with Tine
Biomarin investing in the two biotechnical by-product companies
Maritex and NutriMarine Life Sciences[15]

We had identified marine ingredients as an interesting area, and we
did some research on this throughout the winter of 2001. It resulted
in contacts with several companies as potential partners, and in the
end we prioritised Maritex, at the time owned by Aarhus Olie, and
this entrepreneurial project NutriMarine Life Sciences, that we helped
establish. There were two types of challenges: first, to get these companies focused enough, as they have been focused on seeing many
different opportunities, and following too many paths, and, because
none of them had any cash flow to build upon, something had to be
found that we thought we could realise, to get a sound development.
(Per Magnus Mæhle, Tine)

This was a new kind of industry, different from the fresh fish in
Marian, but with some similarities with other activities, in particular as
found in Tine Ingredients. First, there are product optimising activities
involving utilising as much of the raw material as possible, including
the idea of creating economic value from by-products (or 'waste'), as
ingredients in various industries, related to food, health and cosmetics.
Second, it involved doing business with industrial customers, collaborating closely to meet each customer's needs. However, the coordination effect between these companies failed, and interaction with Tine's

resources proved to be difficult in these ventures. Altogether, it has been a hard and bumpy ride, learning both about the biomarine sector and especially about what Tine's resources can be used for in this setting. Still, Mæhle was hesitant to judge any part of the process as failure. Admitting the bad luck of investing in INAQ just prior to the 2000–3 downturn in the aquaculture industry, causing big economic losses before pulling out, Mæhle still argued that all the other 'should still be seen as invested', meaning that the outcomes of these ventures were yet to be seen. However, NutriMarine Life Sciences was sold after collaboration problems with the entrepreneurs/scientists who had started the company, and who held all the knowledge in the firm. Maritex had huge initial problems both with choosing a direction, starting out with almost 50 different products (from cod stomachs to collagen peptides), and with delivering on technological and quality premises. It would take Tine seven years, large investments and huge efforts to develop Maritex into a high-quality omega 3 refinery, before, in 2008, they could launch the first dairy and bakery products with added marine omega 3. It is always hard to argue for innovation projects during their development, as when the business newspaper *Dagens Næringsliv* included a critical article[16] about Tine's 'failed' biomarine projects. When innovations meet such resistance from internal and external actors, they depend on convincing representatives to maintain support. Mæhle was one of these advocates of the biomarine strategy, as he did not give up as a result of having to face a few obstacles.

Restructuring Tine Biomarin

Obviously, different opinions of a blue-green innovation strategy were represented in the management and board of directors at Tine. As time went on, these differences became clearer as there was an impatience for results, and different attitudes towards the uncertainty of these projects:

> Yes, at Tine's board, there are different attitudes on whether it was smart to enter into these marine projects. But so far, the board has been unified about the decision that has been made, but there are different attitudes about the degree of patience, and on whether we have chosen the right level, and if we have chosen the right projects. (Hanne Refsholt, Tine)

According to Jan Ove Tryggestad (board of directors at Tine), NutriMarine Life Sciences became too vague to continue working with them, Maritex had to take on big economic losses due to a costly and time-consuming upgrading of technology, and the Marian project failed to build on realities – that is, they had problems with suppliers and lack of market access. Then they had turbulent times, with changing management in the companies. When Gunnar Hovland took over as managing director of Tine Biomarin in November 2002, and later continued as director after the merger with Tine Ingredients, he sought to clarify the principles for doing marine business at Tine. Freshness as the main advantage in relation to international competitors, synergies with their core competence and a certain scale of business were, in his view, necessary conditions for succeeding with biomarine activities at Tine. His conception of the biomarine strategy was clearly based on the responsibility of moving these projects closer to commercialisation, feeling the pressure for results from a heterogeneous portfolio of biomarine projects:

> Job number one for me, in all the companies, was really to clean up. Because it was an incredible chaos, both in the strategic platform for why Tine was in it at all, it was anchored to an illusion and not in realities. And there was chaos in the companies as well, they were all entrepreneurial ventures, with entrepreneurial culture and entrepreneurial mess all over, with neither focus nor progress in the important areas for business. (Gunnar Hovland, Tine Ingredients)

When it was decided by the Tine board of directors to sell NutriMarine Life Sciences, we can read the argumentation used for suggesting a decision in an excerpt from the board meeting case papers:

> Orientation about NutriMarine Life Science: Tine chose NMLS as partner based on expected synergies with our product development and sales of special nutritients. After two years of work, the company has become more oriented towards pharmaceutical industry, and it is difficult to achieve synergies with Tine. (Tine board meeting case papers, 7 April 2004)

The change of strategic orientation within NMLS was here used as the reason for why they had not achieved the expected synergies with Tine. Furthermore, the need for more resources – on financing and marketing the venture – served as arguments for selling their shares in NMLS. In the board meeting,[17] the management argued that the process related to clinical documentation (in pharmacy) was expensive and uncertain, that it would take at least another year before anything would be ready for the market, and that the venture was far behind the business plan schedule. In addition, by defining the project as 'outside' Tine's business areas, an argument was made for taking a loss from selling the shares rather than losing even more on continuing with this venture. Several of the board members expressed that some of Tine's owners were increasingly sceptical and impatient regarding the economic outcome of the blue-green projects. Some other board members expressed support for the management's analysis and suggestions, emphasising also the collaboration challenges between NMLS and Tine. The board decided in unison to sell the shares in NMLS. It is clear that it had reached a point at which top management asked fundamental questions about the blue-green strategy, putting Tine Biomarin under pressure. The process had taken more time and money, and been more difficult to handle than was expected, leading to a call for evaluating efforts, and for improving economic results. They found it increasingly difficult to defend these priorities towards the owners – farmers supplying milk. Lack of competence, both on the board and in managing the projects, was seen as the main problem. Discovering the gap between initial pictures and the reality, and 'tidying up' the resulting mess, had been a learning process that hopefully would improve their ability to succeed in the future. Reviewing the biomarine strategy, Hovland emphasised marketing, internal capacity and competence, partner interests and process control as the main challenges. In being left with only two projects after the process of 'tidying up' in Tine Biomarin, Umi No Kami and Maritex, the quest for commercial results in these ventures had to be intensified. In particular, Maritex's business model of producing and marketing ingredients to industrial customers became an argument for merging Tine Biomarin and Tine Ingredients. By this, Tine Biomarin's position as a hub between R&D and commerce had been moved into a clearer commercial position, though still dealing with advanced technical issues, like:

> getting the oil into dairy applications, and sell it both nationally as a demo and internationally as application. That business logic was pretty similar to what we do with dairy ingredients. In a way, you have to do relatively heavy work on the research part, the application part, and the knowledge part to get an ingredient to be stable in a product. And then we sell not the oil, but the solution. (Gunnar Hovland, Tine Ingredients)

According to this argument, it is less clear how well the fish sausage of Umi No Kami fitted within Tine Ingredients. In this case, it was more a case of maintaining control of the project by those who had interests and ideas about Umi No Kami's further development. The process of getting the collaboration between Maritex and professional groups at Tine took time:

> Maritex has had a culture of fixing everything themselves, like a home-fixing culture on everything, and R&D has been there and helped them on several critical issues, but the cultural difference seems so different that we have struggled to make our R&D a natural part of Maritex. (Gunnar Hovland, Tine Ingredients)

Thus, not only geographic differences, but also 'cultural distance', meaning differences in professional and industrial practices, made it difficult to get the collaboration on track. In the communication of both parties I noticed a total agreement to draw on Tine's resources with regard to technology, marketing and economy, in order to develop and commercialise Maritex' business. Yet, in practice, I could not see much of this;[18] there were some attempts related to technology development and to production routines on hygiene, but these did not have much success – not until Maritex's management was changed to include people from the dairy industry did matters improve (2005/6). Again, the issue of 'quality', or nutritional standards, as defined by Tine, was highlighted as the main technological – and thereby also a commercial – problem:

> Maritex has had terrible results, and nearly went broke. They have not succeeded in the market. This is partly because the ingredients industry must have nutritional standards in their production, and the marine sector has not come as far as in agriculture. (Gunnar Hovland, Tine Ingredients)

Obviously, this has to do with having challenges on the marketing side, in being an entrepreneurial company without many initial relations to their markets (industrial customers), and with more than 50 products to sell with almost no resources in sales and marketing. To say that it was unfocused is perhaps an understatement. Another issue is that it took Tine, as owner, several years to realise this as a problem, and then radically restructure the company.

It had been a demanding challenge for the Dairy Cooperative to develop a blue-green innovation practice out of the new strategy, both with regard to the technological aspects (that is, quality standards) and on marketing (focus on customer demand). This led to more spending of financial and human resources than expected and hence created frustration and pressure for results in the mother company. In the case of Maritex, Tine Biomarin ended up buying out the other owners of the fish oil company to ensure economic and managerial control of the company. After lengthy and expensive upgrading of the refinery, and brutal restructuration shifting out management in two rounds, their omega 3 could finally be tested commercially as an ingredient in Tine's yoghurt and in Goman's 'Coast Bread'. It could also be launched in the international industrial market for functional foods ingredients, but it is too early to tell whether it will succeed at the time of writing this book.

As most of the biomarine projects in Tine have been of international dimensions and ambitions, the question of intellectual property rights (IPR) has come to the fore. The idea of producing 'value-added' products from biomarine resources is clearly based on two strategic factors: first, the access to fresh and superb raw materials; and, second, the competence of the Norwegian food sector. Without taking advantage of the freshness and developing unique fields of expertise, labour will be cheaper in other countries and regions, and consequently it will be more profitable to do processing there.

> The only thing that is cheap in Norway is competence. Everything else is very expensive. And then, to put that competence into our products and solutions, get it up, get patents, protect our IPR, this is the way forwards for us. Our products, our milk, are without value. In an international setting, we won't have a chance. (Gunnar Hovland, Tine Ingredients)

This is a strong, but probably qualified statement about how many of Tine's present products will come under pressure from international competitors. Their complex logistics and distribution system, quality control and market relations will not produce the same advantages when trying to compete with international competitors buying milk from far more efficient milk farms. Without a serious restructuration of the Norwegian milk farms, Tine might be outcompeted on price (Olsen and Espelien, 2003), since consumer milk is a commodity, without much potential for building and protecting knowledge-based advantages. Knowledge-based strategies have their potential in more processed food products and ingredients, such as cheese, fermented milk products, ready-made foods, etc. The situation in aquaculture is similar. With regard to the production of raw materials Norway still is very competitive; hence, product development from fresh fish might become a profitable area. The processing of frozen fish, on the other hand, can be done anywhere in the world, as the costs of labour weighs more than the costs of transport, and, in the EU, customs tariffs are lower for unprocessed fish. Strategically, two questions are then of importance for Tine. What kind of knowledge is useful in their biomarine activities, and how can they take and protect a position there? Tine seems to have learned some lessons on this:

> It is about competence on nutritional standards, which does not exist at all on the marine side. They will never manage to industrialise food production on their own. It is about picking the raw material apart and putting it back together in an appropriate way for consumers and industrial production. (Per Olav Skjervold, Tine R&D)

At this point in time, Skjervold had left Tine for a position as director of innovation in one of the major fish feed producers (EWOS Innovation). Where Brendehaug (director of research, Tine R&D) and others said early on that they would attain competence on fish in collaboration with fish actors, Skjervold saw a shift *within* Tine towards valuing their own competence higher. In spite of his critical attitude to the early organising of Tine's biomarine activities, Hovland (managing director, Tine Ingredients) admitted that it is a learning process marked by uncertainty, with new and commercialisable knowledge as the aim:

> *Researcher*: So, you say that this is an attempt at trial and error learning?
>
> *Hovland*: Yes, in practice it is. And the theory is then to look for unique advantages, and add our competence to these unique advantages, and try to protect them in some way. Whether via branding and trademarks, or patents, or whatever, and then test it in the [international] market.

Going down that road of finding 'unique advantages' in the blue-green cross-section had not been done before by industrial actors; hence, the identification of opportunities, whether based on market demand or technological possibility, had to be exploratory, and not as rational as corporate management accounts suggested.

When confronted with the status of Tine's competence on IPR, Hovland was very clear about their shortcomings: 'If we are honest with ourselves, we don't know much.' Having met the demands for IPR in meetings with potential international industrial customers in some of the biomarine projects, they realised the importance of developing knowledge and routines about this in Tine Ingredients. Activities for upgrading Tine's competence on the issue were started. Yet from starting to work more seriously with IPR to having included it as an integrated perspective via strategies, routines and competence in the entire organisation takes time. IPR can also be viewed as a way of making knowledge exchangeable, and providing negotiation power in relation to large international business customers:

> The drive for increased focus on IPR at Tine is not mainly related to the fish ventures, even though they are a part of it, because they are international, and the patenting process and the contract dispute on the fish salami were a bit deterrent. We have clear signals from the US, related to omega 3, that patents will increase the market value and the interest considerably. Moreover, approvals with FDA and GRAS[19] have to be in place. (Bente Mogård, Tine Ingredients/Tine R&D)

Even if the fish industry has been international for ages, the lack of systematic knowledge development when it comes to processing and product development has led to little use of patenting. According to patent lawyers, Stene and Langan, in Zacco (an intellectual property consultancy), there were almost no uses of patents in the fish industry. They further compared it with the oil industry and the poultry industry, where patents are not only used for 'outcompeting rivals, but also to control the value chain'.[20] In the main case of this study – Umi No Kami/Salma – the whole project started with buying a patent application on the fermentation of fish. Later, with establishing relations to Bremnes Seashore, their high-quality raw materials were also based on patented processing technology. However, the economic value patenting provided in this case, is more questionable. I would suggest that other protection and marketing strategies, particularly related to branding, design and continuous innovation, have been of greater importance.

4
Fermenting Fish: Innovation in Practice

In the previous chapter, I sought to provide an introduction and over-view of the case that I investigated, so that the following more detailed descriptions would be easier to comprehend. Also in the previous chapter, I situated the case study within a larger set of interconnected networks that together form what are often referred to as industries, structures, infrastructures, political discourses and institutions and markets. I would not agree that this is the 'context' of the case study as it is often described. Rather, it should be understood as an attempt at 'stretching' the networks beyond what is common within actor-network theory, in order to describe how local practices interact and become intertwined with other practices, sometimes due to their need for simple coordination or for close collaboration across domains, and sometimes in their collaboration towards forming and distributing new practices to the extent that Practice (with a capital P), or what we might call 'macro-practice', emerges. Examples of the latter could be the emer-gence of a new profession, a new category of products or new constella-tions of users creating a new 'market'. Here, I am arguing, along with a number of process and practice theorists (e.g., Latour, 1999b; Law, 2004; Pickering, 1995; Araujo, 1998; Hernes, 2007), that what is global is built from interconnecting local practices and processes. My ethnographic case study is, therefore, just a small slice of what was going on in the network of interconnected practices during the time period in which I followed some of them. The development of a 'blue-green' innovation programme within Norwegian science policy was another instance of such 'innovation-process'. It was no less local than the fish salami, but it was brought forth by other actors with other obligations and interests, such as managers and bureaucrats of universities, research institutes, the Ministry of Education and Research and the ministries of fisheries and of agriculture. The tracing of these specific and local processes in detail would have been an entirely different study, but it would have nevertheless interacted with Tine's development of a biomarine innovation strategy, and with the barriers to commercial innovation within the fish industry. As mentioned previously, I could also have started out with the relationship between Bremnes Seashore and the University of Life Sciences, and traced the processes of technological innovation happening there for years, as well as the subsequent processes of creating economic value from them, while failing to find users willing to pay extra for high-quality products before incidentally coming into contact with Tine. That would have also been an interesting study. Yet that was not the starting point of this project. My starting point was studying innovation processes within agricultural cooperatives. Then I came across the biomarine activities at Tine, and found this to be an excellent object of study for understanding innovation processes that cross boundaries; the uncertainty of the processes was high, and the need to learn and create new knowledge was also high. Thus, the following detailed description of an innovation process is an example of how micro-practices (developing a fish salami) both create and enact macro-practices (industrialisation and market making of fish). This is because macro-practices are built by interconnecting micro-practices, and because the emerging macro-practice and micro-practice mutually shape and influence each other.

Idea and invention: recombining elements

There are certainly several story lines leading to this product development process. One story line began with Tine's surplus of whey, a by-product from cheese production consisting partly of proteins. The desire to make economic use of this idle resource led to various R&D programmes, and one had to do with experimenting with milk proteins as additives to various food products. Another story line comes from Tine's long history of industrial production and distribution of dairy products and, in connection with this, their knowledge development on fatty acids, fermentation and hygiene. More recently, a story line from a totally different setting, that is, Bremnes Seashore and their development of new processing technologies for salmon in collaboration with researchers from the Norwegian University of Life Sciences, is of crucial importance to the present product concept. Yet another story line related to both the agricultural and the aquacultural sectors, and which at various points interacted with these other story lines, is the development of national 'blue-green' research policies and programmes, which tried to stimulate collaborative efforts across the agri- and aqua-cultural sectors. Nevertheless, in emphasising Tine's role in the process, my account of the case needs to start with a professor in aqua- and agriculture, Erik Slinde. In addition, even if it were to be a story about what products came to constitute the concept of Salma, including fresh fillets and other potential products, the story line from Slinde's idea of fermented fish is still difficult to overlook.

Slinde is what I would call an 'entrepreneurial scientist', in commu-nicating self-conscious creativity and energy, and being always ready to throw out opinions about society, research and business. With previous experience from 'both sides', having worked as a scientist and director of research both at the Food Research Institute and at the Institute of Marine Research, he wanted to encourage the product development of fish:

> I suggested that we could start teaching a subject called 'seafood and product development' here at the University of Bergen, and then things worked out so that I did that for three years. Let's take some food technologies, and then apply them to fish, using fish as raw material, and using food processes, and one of the processes that I know really well is production of salami. I thought to myself, ok, we can make a salami out of fish. (Erik Slinde, Institute for Marine Research)

From this experimental recombination of technologies and materials, based on Slinde's expertise, the idea of making salami out of fish, pref-erably from salmon, was tested. It soon became clear that the fat con-tent in salmon would be a main technical challenge, and therefore a mix of red and white fish was seen as necessary:

> The fat content varies a lot in salmon, it varies from 10 to 30%, thus we have to balance this with a white fish, and the most available white fish in Norway is saithe. And moreover, it is extraordinary cheap. Then we thought that this should be a good product, to make salami out of fish, ordinary Norwegian salami. (Erik Slinde, Institute for Marine Research)

Here, we have an early version of the fish salami. On the one hand, it applied technology from ordinary and mundane Norwegian salami and, on the other hand, it was thought of as a competitor and substitute for exclusive products like smoked salmon and 'speke-lax'. These sources of the idea, which simultaneously associate with the mundane and the exclusive, came to follow the innovation process for years, producing both conceptual ambiguity and flexibility. This 'double identity' was also present in the object's technical and economic challenges. First, there was a challenge in stabilising fatty acids; in the beginning a solution was attempted by blending white and red fish. Second, there was a challenge of market segmentation, whether aiming towards the exclusive 'gourmet' segments or a larger market of 'everyday products'. And third, there was the matter of cutting costs – for example, by blending salmon with saithe, which is a much cheaper raw material.

Slinde was not the first person to think of the potential combinatory opportunities between agri- and aquaculture. Between policy-makers and research institutions, I have described some of the debate on the potential synergies between 'blue' and 'green' innovation in Chapter 3. But a fish sausage – a salami of fish? 'What a bad idea,' a lot of people said throughout this process. 'I only work on bad ideas,' Slinde responded stubbornly, 'because, if something is considered a "good" idea, it has certainly been tried before. In reality, if ideas become good or bad, it is the result of the project, not its cause' (Erik Slinde, Institute for Marine Research). Thus, the idea was born, some initial funding acquired and shortly a laboratory at the Food Research Institute was hired to test it out in practice:

> It costs 60,000 NOK to do an experiment at The Norwegian Food Research Institute. I went there, and got hold of salmon and saithe, and we made mixes downwards, with 100% salmon, 50% salmon, 25% salmon, etc, and I think we did 4 productions. We saw that in the products that contained much salmon, the fat just flowed right out. Then it was back again [to Bergen], in way going back and saying that this was really a failure, we can't make it. (Erik Slinde, Institute for Marine Research)

The first experiments did not go very well. Yet, even though he thought of the experiment as a failure, Slinde still brought the results back to his fellows at ForInnova (the University of Bergen's Technology Transfer Office) and in the Renew-programme:[1]

> I went back home, and came with these nice packages, right, a little like 'dress up the bride'. And then, the economists at ForInnova went down to the store and bought a few beers, and these guys ate it, and

said it tasted delicious. So, I thought that, if three economists are sitting here telling that this is good stuff, then I am sure I can make it better. (Erik Slinde, Institute for Marine Research)

Slinde's interpretation of the situation, even in hindsight, was ambiguous. On the one hand, the first experiments failed, and he was really heading home without positive results. Still, just to be able to show *something*, he 'dressed up' the fish salami and let the businessmen have a taste. Surprised then by the fact that they got excited about it as well as its potential as a commercial product, Slinde realised that if a group of businessmen liked it, it could become something more that just a researcher's fascination about a new technological combination. I was later told by one of the Renew-programme representatives that he thought the product tasted 'awful', but that they thought it was a fascinating project, and some of them had a good relationship to Slinde. Thus, it was decided to continue with further experiments, and start developing a business plan for the project. A bit later, samples of the emerging product were presented at an aquacultural fair, and the visitors were asked what they thought about it. Slinde recalled particularly one episode that, according to him, strengthened his belief in a market for such a product:

Some said that they would absolutely not eat it, but, then it happened. A mother came with a little boy. The boy saw that we had sausage, and he wanted a bite, then he ate it, and asked 'can I have some more?' And you know, then these guys in ForInnova understood that getting kids to eat salmon as sausage, giving them omega 3, then you have the mothers, so then there was a market. (Erik Slinde, Institute for Marine Research)

Selling 'easy omega 3' for children, which is both a new way to approach eating fish, and possibly a market for 'family-friendly fish', this event seems to have provided some reassurance for going further with the project. The next step was to patent the invention, and patenting was judged as a crucial point of the decision: 'If you patent something, you start a commercial race, you have to sell, get someone interested in taking over' (Erik Slinde, Institute for Marine Research).

Step 1 was easy – exploring the idea in a laboratory and then having public funding was a green line to move forwards. Step 2, starting the patenting process, however, would lead to Step 3, seeking to industrialise the project. The great risk, as well as the costs and commitment needed, were considerable, and Slinde and his partners in the TTO could not take on the costs of patenting without thinking about commercialisation right away. Slinde and ForInnova decided to take the next step and write the patent application, thus also going for commercialisation – preferably by selling the patent to a capable industrial actor.

Neptun: research and technology development

Still, how did an agricultural cooperative like Tine come into the picture? Tine was one of a very small set of domestic actors having the capacity to conduct heavy product development and commercialisation processes in the food industry. Yet the emergent cooperation between Slinde and Tine R&D came about through incidental events, and was developed in a common research project with public funding.

During this early period, the participants focused mainly on technical problems related to texture, durability and colour. They also worked on other problems related to the micro-biological quality of available raw materials. In particular, fish coming from the catch side, or white fish such as cod and saithe, did not meet the standards perceived necessary to make the technology work, and they found it hard to change the suppliers' practices according to their demands. This was a dilemma for the innovators: change the practice of fishermen and fisheries to fit their need for 'high-quality' raw materials of white fish, or face the technical and economic challenge of using only salmon in the recipe? With the emergence of salmon farming, it had become considerably easier to control such aspects, but it was at the time considered to be too technologically difficult and too expensive to develop a pure salmon product. This image of the innovation as based on a mix of red and white fish became also anchored to, and later reinforced by, corporate management at Tine.

Collaboration in the laboratory

The fatty acids in the fish sausage would not stabilise thoroughly, as fish fats are more liquid and oxidise (get harsh) faster than fatty acids in meat. Hence, Slinde needed something that could deal with the texture of the product. His quantitative analysis indicated that adding some kind of protein could possibly help stabilise the product. While eating lunch in the canteen, during a day he was working in the laboratory at the Food Research Institute, Slinde came to talk with Berit Nordvi, a researcher from Tine R&D:

> Berit Nordvi was working on milk proteins. Being somewhat idealistic, I thought that if we can use a Norwegian source of proteins it would be better than soy from the US, so I thought, ok, milk proteins are ok, let's test it. (Erik Slinde, Institute for Marine Research)

Nordvi told Slinde about her project on exploring various applications of whey proteins and other milk proteins. According to an industrial logic of 'product optimisation', Tine is impelled to exploit as many of their raw materials as possible for economic gains – in this case, a large surplus of whey, a by-product from cheese production. Tine R&D joined Slinde's project without any rights to the invention, as their interest in participating in the project was partly about the prospect of selling whey proteins if Slinde succeeded, and partly about learning more about the use of this dairy ingredient. Slinde, as a result, got access not just to proteins, but also to both material and knowledge resources on a wide range of aspects of food production. It was partly accidental that Nordvi was present at the Norwegian Food Research Institute that day and that her project fitted nicely with Slinde's present needs for a protein supplier.

This was one of several starting points for what came to be a much larger programme at Tine than they could foresee at the time, a scientific and economic exploration of a number of issues that they hoped would eventually lead to commercially beneficial knowledge, technology and products. During this first phase, the most pressing issue was understood to be to stabilise the rather fluid fatty acids in fish, a rather unexplored issue within food science:

> The fat slipped out. One couldn't stabilise the fat, and so his focus was that they needed to add more proteins. We then did some experiments together, and I was surprised when the taste of the product became that good, and that its taste kept so well over time. From my viewpoint having worked with oxidation of fish oils and things like that, this was not possible. And then, suddenly, I was caught by the project, it was terribly interesting. (Berit Nordvi, Tine R&D)

The effects of interaction between fish fat and milk proteins surprised Nordvi, and triggered interest, as she was 'caught by the project'. There was a short way from the triggered curiosity to organising a common project. Thus, after having found this interesting epistemic object, how did the researchers mobilise the considerable resources necessary to work with it? It seems that a combination of good relations with the Research Council and effective argumentation towards the Tine management did the job.

Funding the Neptun project

Despite being driven by a fascination and passion for the unknown, industrial research is often a practical, almost mundane set of practices, with projects often growing out of concrete problems or challenges. Researchers seeing opportunities in expressed problems related to the organisation of its resources, production and marketing. At Tine, surplus of whey had, as mentioned previously, been considered to be a problem for quite a while; the search was on to find products that could, either for reasons of production or those of markets, utilise a significant amount of this by-product of cheese production, rather than the small amount then used as animal feed:

> Whey is a byproduct from production of white cheeses. In 1999, almost 80% of the whey volume of a total of 840 million litres was used as animal feed. This has not been a very economic utilisation of valuable components of the raw milk material. TINE has therefore during the last decade prioritised building core competence on whey, through research activities on the development of tailored whey protein concentrates for use in the food industry. (Attachment no.3 in the RCN application for the Neptun project, March 2001)

In January 2001, Nordvi and Tine, in collaboration with Slinde and the Norwegian Food Research Institute, agreed to apply for funding from the Research Council of Norway (RCN) for three research projects on protein applications. One of the applications was accepted, and the Neptun project was established. Dialogue with actors in the RCN helped them to write the applications in line with strategic priorities for research funding. Getting approval for a research project like this was a matter of satisfying several stakeholders. After having received support from the Tine R&D management, via argumentation for learning about the application of whey proteins, the next gatekeepers were corporate management at Tine and the Research Council, both of which had somewhat diverging interests, hence making this an exercise in political sensitivity when formulating the project proposal:

> We sent the application, and it was framed in a rather unusual way, in dialogue with the Research Council. It was out of the question for

Tine to front fish in that period and so we quite simply fronted the ingredients side. (Berit Nordvi, Tine R&D)

From the point of view of Tine's corporate management, the timing was not right to become an industrial representative for the public 'blue-green' innovation programme. Their biomarine innovation strategy was not yet ready, and also awaited approval from the board of directors. Nevertheless, in having developed some kind of basic competence on protein applications both on products from meat and fish in earlier projects, the concept of fermenting fish revealed new and unexpected functions of milk proteins which opened up a new angle of research. In the project description of Neptun, this was connected to an even longer tradition forming the background for curing and fermenting salmon: 'In Norway today, we produce smoked and graved[2] salmon. Smoked salmon has a relatively high content of water, while gravlax is characterised by a strong fermented flavour' (Project description, 'Salmoni', March 2001).

For a company having a strong identity in traditional rural culture, and for a domestic research council with responsibilities for stimulating national and regional industry, associations to tradition and popular products in the Norwegian cuisine might have contributed to making this project relevant. Moreover, 'Neptun' was, without a doubt, a word with marine connotations, associating the project with the initial spark from Slinde and his sausage. Still, Nordvi explained the title and the project intentions in somewhat ambiguous terms:

This is really not a project just about marine applications of milk proteins; rather, it is a project of the general applicability of milk proteins, of whey in particular. But since aquaculture was a hot topic for research funding at the time, we designed the project application with that in mind. (Berit Nordvi, Tine R&D)

The important aspect of the project from Tine R&D's point of view could not have been fish at the time, nor was fish Nordvi's personal and professional interest. It was the opportunity to exploit whey in new applications, hence both producing an interesting research problem and making business out of an existing and problematic by-product, an attempt at reframing whey from waste to valuable resource. The connection to fish, was here argued to be a pragmatic matter, of positioning the project where funding was available. At the same time, Slinde's fish salami played an important role both in initiating the Neptun project in the first place, and in serving as an example within the project application of the implications of the study. These efforts at strategic navigation indicated tensions and a need for political caution, and care in avoiding politicising the project. From the project application to the RCN, the fermented fish nevertheless received a privileged position as part of both the main and subsidiary objectives of the project:

Main objectives of the project:

Produce at least one tailored whey protein concentrate (WPC) in pilot scale that is characterised as very well suited for use in fermented and dried seafood products.

Subsidiary goals:

1. Produce and characterise tailored whey protein variants...that are assumed to be well suited for use in seafood products....
2. Acquire a quantitative and reproducible method for evaluating different WPC variants in a dried, smoked and fermented seafood product model.
3. Test new tailored whey protein concentrates...in this fermented seafood product model.
4. Ensure the completion of a PhD...related to emulsion and binding of unsaturated fatty acids from fish in concentrates and in fermented seafood products.
   (The RCN application for the Neptun project, March 2001)

Thus, in practice, this project was designed as experimentation with, testing and designing – specifically – protein concentrates suited for the fish salami. Yet, the research problems to be dealt with were framed in more general terms, specifying the micro-biological research questions rather than their possible practical applications, though with a strong association to fish: 'In terms of research, this implies that it is important to understand what mechanisms are important for binding the unsaturated fat in fish products' (The RCN application for the Neptun project, March 2001).

Finding new mechanisms for binding fat from fish, and thus learning how to strengthen and control them, was the task at hand, as formulated in scientific terms. Further, this was innovative; indeed, the problems of fluid and oxidising fatty acids in oily fish, related to fermentation, was an unresolved – and probably rather unexplored – issue at the time. Thus, while Tine had a great deal of knowledge about their milk proteins, or how to make concentrates of them, for use in milk-, meat- and fish-based applications, the challenge of mastering these difficult problems – of controlling and curing fresh and oily fish – was new and exciting, and opened up for new ideas of industrial food production.

With funding from RCN, the project gained some size in terms of the potential actors involved. Some of the monetary support was intended for a PhD stipend and, after thinking it over, Nordvi decided to take the opportunity herself. As part of the early documentation, a market report on fermented sausages was written, seeking to back up the claim in the project proposal for the commercial potential of a project on whey proteins:

> Total turnover for fermented sausage in Norway is equal to about 1.83 billion NOK a year.
>
> | | |
> |---|---|
> | 1% of total turnover: | 18 million NOK a year |
> | 5% of total turnover: | 91.5 million NOK a year |
> | 10% of total turnover: | 183 million NOK a year |
>
> Total sales value of fermented sausages in Europe is equal to around 194 billion NOK a year.
>
> | | |
> |---|---|
> | 1% of total turnover: | 1.9 billion NOK a year |
> | 5% of total turnover: | 9.7 billion NOK a year |
> | 10% of total turnover: | 19 billion NOK a year |
>
> (Market estimate for the Neptun project, on fermented sausages, 2001, Tine R&D)

This qualifies as a very early evaluation of market potential. Notice how the argumentation is pointed towards the effort of commercialising the *sausage* (which was still Slinde's property, not Tine's), and not the proteins per se. Further, the market was described in terms of 'total turnover', not of cured fish products, but of salami – fermented meat; hence the outlined market potential became enormous, and if anyone in addition had ambitions to take on a considerable share (1, 5 or 10 per cent) of this, then this would, of course, become hugely profitable. This is very similar to a large number of 'market reports' in most business organisations, almost like a ritual. What do such 'market reports' with estimated 'market shares' do when the product is not finished and its use and its end-users are not, and cannot be, known? Are they necessary to produce convincing argumentation? Or is it just an obligatory and meaningless part of the business-plan ritual? And would anyone believe such a story? Later, several times during the start-up of the UNK project (after buying the patent application from Slinde), and when Salma went on its first marketing tour, leading figures in the project and corporate management at Tine, told the story of making the fish salami 'the next Jarlsberg',[3] – that is,a big international success:

> Tine finds participating in the further development of the Norwegian food industry to be an important role, especially in the areas where Norway has a natural advantage and with products that may be of commercial significance for our milk based ingredients. (Attachment to the RCN application for the Neptun project, 2001)

The research was in this way connected to the Norwegian food industry more generally, and to Tine as an actor with certain responsibilities in developing this industry. In one and the same paragraph, ideas about sectoral unity, food markets, Norwegian industrial identity and Tine's future income are reinforced in relation to each other. Note also that when 'commercial significance' is used in this setting, it has to be viewed in relation to Tine's total turnover of more than 11 billion NOK (2001), thus strengthening the impression of high ambitions in the project.

The aim of building up competence on protein applications was present in the project throughout this phase. In *practice*, as a genuine interest of the involved project participants, and in *discourse*, perhaps, as a backup in case the commercialisation of fish (the fish salami) did not work out. By breaking down the composition of the object to a micro-biological level, not as a chef with ingredients of essentially natural origins, but as a biologist, they performed the role of the scientist who, with microscopic precision, separates and recombines material components to arrive at new properties:

> The properties of a fermented salmon product are the result of proteins, fat and water. In addition, antioxidants need to be added to avoid oxidation of the fatty acids. When salmon is used as raw material, the colour in the final product is a result of the amount of astaxanthin in the fish. The growth of bacteria is steered by the amount of added sugar. (Attachment to the RCN application for the Neptun project, 2001)

In this résumé of the process and recipe of the fermented sausage, the proteins were claimed to be the intermediary between fish and meat, between fermented fish sausage and salami from meat, and furthermore: 'These will be seafood products that may become a real alternative to fermented sausages of meat' (Attachment to the RCN application for the Neptun project, 2001).

Thus, it was argued that it was possible to make fish a 'real alternative' to meat, a science-based simulation. Yet something was missing: 'There is need for more basic knowledge of how proteins bind the fluid unsaturated fatty acids from fish at low temperatures and low pH' (Attachment to the RCN application for the Neptun project, 2001).

The relation between proteins and fat needed to be developed, as the potential effect of combining them was not yet fully exploited, nor fully known. Hence, Tine's research capacity and competence was needed, and not only on proteins: 'Tine Norwegian Dairies have their core competence within the production of fermented dairy products. It is not unexpected that fermentation of fish is considered to be a strategic business area' (Attachment to the RCN application for the Neptun project, 2001).

Bacteria cultures were also an area where Tine R&D could contribute with their experience and expertise. Proteins and fat stabilisation were given a more central position in the application than the 'core technology' of fish salami – fermentation. Thus, though almost absent from the formal research agenda of the project, fermentation was mentioned only briefly, in order to strengthen the argument for Tine's involvement in this biomarine project, drawing connections between generic (representations of) biotechnical knowledge and specific research problems in a totally new area for the organisation. Thus, we see how the project team made connections between the tiny micro-biological components of scientists' materials, an innovative seafood product, Tine's core competence and conceptions of potential markets. Let us now move a little closer to their practice of biotechnical science.

Testing different types of fish

The focal problem in research practice during this phase was how to make a tailored protein concentrate for fat stabilisation:

> After this initial work, we chose to use a standardised and dried seafood product model in our further testing. This so-called 'Neptun-model' is based on a combination of salmon and saithe, and can be viewed as the prototype 1 of a specific fish product. To build knowledge on how different protein-containing ingredients influence different texture properties and fat binding in this product, a wide

selection of commercial protein-based ingredients was first tested. Afterwards, eight different tailored whey protein variants were tested in the model. (Final report on the Neptun project, 2005)

In this scientific setup, an initial model on which to test and measure the research results was made, a 'quantitative seafood model', that would make statistical analysis possible. Later in the project, another model was made, a physical prototype. Unsurprisingly, it was not made from scratch; it was a standardised prototype of the fish salami – a conceptually and materially 'locked' version. The quantitative model (here, referred to as model 1), had subsequently become a material prototype (referred to as model 2). In order to run practical experiments with various proteins and ingredients, and then measure and evaluate the results, it was necessary from a scientific point of view to decide on a specific version of the fish salami recipe. This was an act of 'locking the object', or temporarily stabilising it. In this way, models 1 (theoretical/statistical) and 2 (physical prototype) could be maintained in a relatively stable relation, representing both each other and the *idea* of a 'fish salami'. The results of experimenting with the physical prototype (via a controlled variation of certain ingredients) could be fed into the quantitative model to produce scientific knowledge on the effects of different proteins on fermented fish. Thereafter, one could decide on the most suitable version, and then jump back to the prototype using the better technology. The original idea, to combine white and red fish, was maintained in the patent application, the Research Council of Norway project application and in 'prototype 1'.

In the Neptun project, a large number of different recipes were tested. Cod, saithe, various Asian species, haddock, mackerel and other white fish species were blended with salmon in varying degrees, along with a few attempts at a pure salmon product. Several of the white and red fish blends produced good results, both on taste and texture, and they found it easier to control the stabilisation of fat in such blends than in pure salmon variants, which tended to have considerable problems of fat flowing out in spite of the added proteins. In addition, some white fish species, like saithe, were a lot cheaper as raw material. A mix of red and white fish was thus regarded as being both technologically and economically ideal. On the other hand, the mixed recipes brought about some problems. First, the colour was not considered delicate enough; it lost intensity from the inclusion of white fish, and saithe in particular made the product almost grey. A solution to this problem was sought by adding artificial colours. A few alternatives of colour were tested, which again represented new challenges in food production, including food regulations allowing some ingredients and restricting others. In addition, there were challenges of image – for example, added ingredients made the product seem less 'pure' and 'natural'. After a while, the adding of colour was dropped completely. Here, salmon provided the 'model colour'[4] for the other recipes, a delicate and clear red/pink, both signalling its salmon content and its salami identity. Second, it turned out to be difficult to get deliveries of white fish that were treated with the necessary hygienic standards – that is, with a low enough content of bacteria. This was regarded by some as a problem that was almost impossible to overcome, due to the general standards of the fish industry, especially on the catch side. Yet, by others, it was regarded as something that could be overcome through working with suppliers, in a combination of making clear demands and teaching, based on the dairy's own expertise on the issue. According to Slinde, fishermen are 'trampling on the fish':[5]

> And then there is the problem, and this is why I was so cross with Norwegian fish industry too, that when we bought so-called saithe, it always had a bacteriological quality that was unacceptable. (Erik Slinde, Institute for Marine Research)

In commenting on the technoscientific competence of Tine, professor Slinde argued, based on his experience with research and evaluation/control tasks in the fish industry, that it was wise for the dairy cooperative to opt for biomarine activities:

> It is appropriate for Tine to go for fish, because of their competence, they have special competence on fragile foods, and we want to sell sushi and sashimi, which are to have less than 10,000, preferably 100,000 bacteria per gram. The Norwegian regulations on nutrition accept 1,000,000 bacteria per gram in fish. We don't accept more than 100,000 in poultry and other food products, but in fish we have an extra high limit, because it is so damned poor. I usually say that 'you should normally not tread where you keep food', but if you look at a trawler deck and look at the guys walking around there, because a trawler deck is not disinfected. A fish that fell on the floor is actually spoiled. And if a piece of meat fell on the floor in a meat facility, it would not get back on the table, but in the fish industry this could happen. (Erik Slinde, Institute for Marine Research)

This diverging practice between fish and agricultural industry was seen as a source of problems by the innovators at different moments throughout the process. For several reasons, this problem – or controversy between fishery and agricultural practices – of micro-biological quality would not be solved before much later in the life of the fish salami. Nonetheless, the Neptun project went on and, gradually, Slinde and the Neptun team started thinking about the commercialisation of this new technology. A patent application had been filed by Slinde, and gradually he started working towards selling the idea to an industrial actor.

Umi No Kami: from science to product development

It is one thing to collaborate on researching a technological problem; it is another to buy the entire idea, taking all the responsibility and risk for industrialising and commercialising the new technology. In this section, I will describe how Tine came to buy the idea of fermented fish from Slinde and the subsequent process of setting up a product development project, Umi No Kami ('the god of the sea' in Japanese), aiming at developing recipes and a commercial concept for the innovation. Below, I describe some of the work on technological challenges, how the project group sought to market information about such a product, and their efforts at conceptualising it, through testing bacteria cultures, proteins, additional flavours and colour. It was a process of asking experts (marketers, chefs, etc.) and consumers in different geographical areas about a product that was not finished and no one had heard of before. In this phase, they worked almost exclusively with mixed recipes (red and white fish). In addition to the directions of the initial idea and the patent, this was also a management-imposed boundary that hindered the exploration of other directions. In spite of aiming for commercialisation, the multiplicity of choice in this research-based setting did not seem to narrow; instead the objects under investigation seemed to have steadily expanded during this period. The Umi No Kami project went on for more than two years from its start in 2002, and it seemed that the number of possible routes, providing different sets of opportunities and problems, continually increased. Lab work was able to explore and develop the object into a product that was stable in production in a number of various ways, but less able to choose among alternatives.

Filing and marketing a patent application

After having filed a patent application, Slinde's aim was to sell the idea to a large industrial food producer, which in his view was the most realistic way for such a product to make it into business. Along with a technology transfer office, ForInnova, Slinde started presenting his invention at national and international food fairs. It should be noted, however, that neither the shape, taste, nor the concept of this object weas very developed at the time. Thus, they presented a set of different versions, or combinations of fish types resulting in different colours and tastes. No naming, packaging, segmentation, categorisation or other market-oriented exercises had been done. Hence, the presented object was in reality an emerging technology with a few potential product versions. After the previously mentioned responses at the Aquanor fair, the effort to sell the idea to an industrial actor increased. Presentations and offers were made to various international actors, and to Tine.

One of the first things Slinde did was to present the patent to Gilde, the major Norwegian producer of meat – both cured and fresh – and perhaps the Norwegian agro-food actor that would have the best qualifications for industrialising and commercialising the product. Yet they were clear on their refusal of the offer; they were not interested. Tine was then left as one of very few capable actors in the country to both develop the product further (R&D) and to commercialise a product falling between established product categories and marketing channels. Due to the ongoing collaboration on the Neptun project, a reason for contacting the corporate management at Tine was already in place, and a first meeting was arranged. In the beginning, even though they were somewhat interested, Tine did not make any decisions. Slinde's other option was to sell the patent internationally – for instance, to a Japanese actor visiting Bergen at the time. He expressed interest, and considered the price, a price significantly higher than what Tine had been offered, to be rather low:

Anyway, it so happened that a Japanese man came, and he wanted to take the product back to Japan, sell it and make some profits. Then he asked what we wanted for it, and we answered '20 million Norwegian kroner', and then he responded, 'ah, very cheap!' (Erik Slinde, Institute for Marine Research)

Being experienced with fish products, the Japanese man took away both the bread and the salad that accompanied the fish salami, when tasting the product:

And he, right, we always served the sausage with bread and some salad, because it is a problem, that all food made from fish will become harsh. You don't taste the harshness, that's why you always get cucumber with mackerel. There is some aldehyde in it. Right, and then he picked up the piece of fish, and he ate only the sausage, of course, he was Japanese. (Erik Slinde, Institute for Marine Research)

In hinting at the knowledge about fish embedded in the Japanese culture, Slinde demonstrated the potential of the product, was able to create interest in spite of being at a premature stage of development. With the man still interested in the patent, after tasting a fermented, and slightly oxidised, piece of fish, Slinde went back to Tine for another try to convince them about investing in the project.

I said to them 'please, can you please buy this product, so that it stays in Norway? I think this is important', I said. 'And unfortunately, you have to face that the guys selling it, they understand now that it is worth more', since they have gotten signals that the Japanese want it. (Erik Slinde, Institute for Marine Research)

Thus, the foreign potential buyer proved to be an efficient argument for getting Tine to take action, and within a few days after being confronted with this competitor, Tine had decided to buy the technology. Yet if he could get several times more money for the patent application from the Japanese actor, then why did he sell it to Tine? Slinde had done business from his own research in various ventures several times before, and had some experience with the business aspect. He gave two different explanations for selling it to Tine. First, he considered the offer from Tine to be a safer bet: 'Values are only values when they have reached my bank account,' he said; thus, in his view it was better to accept a secure and quick offer, than engage in long negotiations with an experienced international actor, and take the risk of ending up with no deal:

Yes, well, not just twice as much, but I understood then that we possibly could get five times the price from Japan. I have worked with those huge 'golden bird' projects and I never run after them anymore. I am not interested in them. And this is the issue, Japan was an opportunity, but Tine was a reality. (Erik Slinde, Institute for Marine Research)

Second, as mentioned above, he expressed a wish to contribute to Norwegian food industry, as 'fish is what we are supposed to live on in the future'. Obviously, he wanted to give the product the best possible conditions for development and commercialisation, hence preferring to deal with the most capable actors. In Slinde's view, those who were most capable of industrialising seafood were not found in the fish industry. Agriculture was seen as superior to aquaculture, in scientific, industrial and marketing knowledge.

In order to ensure that not only the piece of paper resembling a patent application, but also the practices, ideas and experiences of which the fish salami was part (i.e., its history) were represented in the project's further process, they hired Slinde as a part time consultant. Still, despite this intention, his availability as a resource for the project organisation remained untapped for large parts of the project. Only in relation to the already established collaboration with Nordvi and her whey protein research was Slinde included and used.

The decision to buy the patent application at Tine was strongly related to the selection of the biomarine sector as a new strategic business area at Tine. Corporate management at Tine became convinced of the image of a fish salami, a 'meat product' based on fish. It fitted nicely into their ongoing revision of strategy, in which, as mentioned in Chapter 3, a new 'blue-green' business development area was under articulation: a strategy between the company's development of new business areas in the face of increasing competition and its consciousness of 'responsibility' in the development of Norwegian food politics. Corporate director Mæhle, central both in developing the biomarine strategy and in establishing the Umi No Kami project, emphasised Tine's relation to the department of agriculture:

> Even if Tine has an important role within the agricultural food industry, as an extension of the state's agriculture politics, it still is very important in relation to the future legitimacy of Tine exactly that we go this broad, for example by doing things in the biomarine area. (Per Magnus Mæhle, Tine)

The fact that a business cooperative would have a designated role to play in present-day national politics, performing certain tasks[6] in the dairy and milk farming sector on behalf of the state, is perhaps counter-intuitive in most western economies, but here we see how a corporate director expressed concerns about Tine's future legitimacy, and how crossing the boundary between the blue and the green domains possibly could benefit their future relations to the state. Despite occasionally referring to the blue-green discourse, I still had the impression from management at Tine that the main reasons for choosing the blue-green path were their own analyses and strategic considerations of business. However, Jan Ove Tryggestad (board member representing the farmer-owners) argued that the 'public discourse on blue-green innovation is the strongest driver for these activities', and that Tine had received 'praise from government ministers' for this. He further emphasised how the quest for political legitimacy and goodwill, together with the allocation of research funding (Research Council of Norway) and the development of aquaculture – that is, fish farming – had provided incentives for going into biomarine activities, thus strengthening a long-term commitment to blue-green industrialisation.

The project carried a promise of a profitable utilisation of Tine's surplus of whey, yet what else was in the patent application? According to Slinde, only imagination set the limits:

> I realised that the technology that Tine received, fermentation technology on fish, in practice means that if they want to develop this area, it is almost like making cheese of milk. And when you know how many cheeses there are, you can imagine how many products you can make of fish. (Erik Slinde, Institute for Marine Research)

In other words, he argued that Tine attained knowledge of a generic technology, with the potential for a wide variety of product variants. Still there are many ways of framing the question, what Tine actually bought from Slinde and ForInnova, depending on who you ask, and, not unimportantly, when. The viable path, according to Slinde, for commercialising such a product would be to go international, as the Norwegian market would be too small. Normally he would have used a patent office in London, but out of a sentiment of idealism in wanting to strengthen the patent competence in Bergen, he used a local patent office. The result was 'not optimal', but still Slinde was 'satisfied' with their job. A couple of opposing claims arose, which created work in getting the patent application approved:

> They received two opposing claims – fantastic! – one from USA, which, in my opinion, had nothing to do with the patent at all, and then some young people in Ålesund, that claimed that this is their invention. There will always be someone arguing that he had come up with it before. Fair enough, but we can disregard this, it is a digression. (Erik Slinde, Institute for Marine Research)

However, in the Tine system, these opposing claims became more than just a digression, both in terms of challenging their knowledge of patent processes and in terms of increased doubt about whether the technology was novel and specific enough for approval at all. From a status report on the Umi No Kami project, we can read the following:

> Norway: Preliminary approval is withdrawn due to the received objection. TINE answers by maintaining our claims as to its novelty and innovativeness. Internationally (PCT-application): International Preliminary Examination Authority concludes that the level of innovativeness is not accepted on any of the aspects, nor on novelty. Still, there is a positive opportunity for approval in some countries: EU, Japan, South-Korea, USA and more. (Status report, Umi No Kami, 2003)

Commenting on the patenting process, with the application still not approved, Bente Mogård, in the Umi No Kami team, described some of the problems:

> The application was almost ready, but it was not good enough to protect us against objections on novelty, innovativeness and industrialisation, thus we have spent a lot of time trying to defend ourselves. But at the same time we have changed the production process, and it is now less related to the patent, hence not as important anymore. (Bente Mogård, Tine Ingredients)

On the one side, it was hard to produce argumentation for the patent to convince patent authorities of its novelty and innovativeness. On the other hand, as the Umi No Kami project moved on, the technology and recipe changed and so the patent perhaps would not be as relevant any more. What were the patent application and the Umi No Kami project about: a fermented fish sausage with similarities to salami? A generic technology having great potential for multiple products? Or a failing application unworthy to be judged by patent authorities as novel and innovative enough? Unsurprisingly, the development of different understandings of this contributed to escalate conflict, as Slinde recalled from Tine Ingredients' attempt at renegotiating the contract with ForInnova:

> When Gunnar Hovland tells me that he will never do business with ForInnova again, well, that matter has two sides. The question is,

what do they think they are buying? And what do they actually buy? And then, I have learned this now, every contract containing payments over time where people don't do anything, in other words, when you licence something, or royalties, or some kind of success fee, these things almost always end up with disputes. (Erik Slinde, Institute for Marine Research)

In addition, the inclusion of a success fee in terms of additional payments from Tine to ForInnova for each new market (i.e., geographical region) in which the fish salami would be introduced was increasingly felt as a burden on the Umi No Kami team, and hence they sought other ways to deal with the issue, such as negotiating changes in the contract, documenting how different the final product was from the original recipe in the patent application, and so forth. The development of fish salami demanded a lot more resources than they first thought:

I do think that Tine thought they bought a sausage that was more or less ready developed, but then it turns out that it is not finished, and that it would take a lot of work. We bought a patent, or, in my way of seeing this, is that we bought a technology, about how to bind fatty acids in fish to enable making of sausage. (Janne Haugdal, Tine R&D)

Despite being less developed, and more generic as a technology, they nevertheless kept on developing the innovation and its potential application. I will now attempt to give a closer look at the initiation and organising of the product development project, Umi No Kami.

Starting the Umi No Kami project

The act of buying the patent represented a shift in focus and organisation of work by forming a new project aimed specifically at the technical and conceptual product development of this 'sea salami' into a commercial product. While the Neptun project had been working on fundamental processing technologies, Umi No Kami was defined as a product development project, taking the initial idea, knowledge and technology into developing recipes, marketing concepts and testing and adjusting the technology to make it stable and under control:

Neptun has continued as a project, and is almost finished these days, and has been purely research related. Umi No Kami was more of a combination of research and product development. My role has been to integrate the research into the product, and then take it to an implementation phase. (Lars Petter Swensen, Tine R&D)

Neptun had for almost a year been preparing the ground for Umi No Kami, by exploring and producing knowledge on the basic technologies used in the sausage. The connection between these projects was maintained during the two first years of Umi No Kami, pooling the resource base for both projects, and supporting each other with technological research questions and possible solutions. To some extent it seems that the activities of Neptun were turned towards serving Umi No Kami. In commenting on the early motivation for the Umi No Kami project, Nordvi expressed a wish to continue working on the invention, beyond the scope of Neptun:

> I was eager to get Tine to buy the patent. Erik Slinde wanted to sell to Tine, he had also tried elsewhere. And I wrote a report together with my leader, who was Ove Johansen, which I sent to Per Magnus Mæhle.[7] He had tasted the product earlier, when we tried to argue why we should get involved in this, and consider buying the patent. So, I tried to give as good of a background as I could about the product, and I got some numbers saying that the international market on fermented sausage from meat are pretty big. (Berit Nordvi, Tine R&D)

This indicates that the mobilisation for buying the patent went through alliance building with central actors in Tine's business development activities, and through producing arguments according to corporate rituals (e.g., market share). The motive for initiating these projects seems ambiguous. On the one hand, the reason for doing Neptun, and relating it to Slinde's patent application, was the opportunity to exploit more of Tine's surplus of whey in an industrialised fish product probably produced by someone else. In this sense, the project looked similar to other projects related to Tine Ingredients' business area – selling various ingredients from milk, and thus often participating in R&D activities to enhance understanding of how to use these raw materials in different settings. But on the other hand, Nordvi revealed also an early hope for taking over the whole thing. Hence, the process of convincing the organisation to get involved with the fish sausage, and to buy the patent, began already with the establishment of Neptun. In part, they started an intentional process of presenting and arguing for this as a golden opportunity. In addition, it seems as if the object itself had some kind of ability to gather interest and enthusiasm at Tine on its own. As soon as people in R&D, management and marketing started to become familiar with the product, various narratives of commercialisation started emerging between the product and its 'interpreters', so to speak. Few were left ignorant of the fish salami; its strangeness and technological wonder ensured curiosity and fascination, as well as scepticism and laughter, in the organisation.

Organising the project

It is noteworthy how Neptun was organised as an open and science-based project involving a number of different people from different organisations, while Umi No Kami, on the other hand, was organised as an internal project at Tine. The Neptun team, although based within Tine R&D, worked together with university researchers as well as people from marketing and consultants with expertise on related products, and with suppliers of fish and technical equipment.[8] Umi No Kami, on the other hand, was organised as an internal project with few relations to outside actors.

To a larger extent, the Umi No Kami project was based on a cross-professional task: to develop and stabilise both the technology and a marketable product concept. Thus, Umi No Kami had to involve scientists, technologists and people with expertise on marketing and design. In addition, it was more strictly tied in with the corporate innovation strategy, hence also involving the management when setting or changing the direction of the project. All this made the project more exposed to potential conflicts and tensions. As a result of Tine's recent change into a corporation and the following reorganisation of management, Karl Inge Rekdal became redundant in his position as managing director of one of the dairy cooperative's regions, and he was then offered Tine Biomarin's Umi No Kami project instead. Even if it had been a long time since he had worked on the shop-floor, his training as sausage maker gave him some professional credibility for leading this innovative hybrid fish/meat/dairy project. From the start, the project participants were Bente Mogård (internationalisation), Hilde Torvanger (market), Berit Nordvi (research), Frode Fimreite (market) and Janne Haugdal (product development), in addition to project manager Karl Inge Rekdal. Then a number of additional people were associated, such as Kjersti Østbø (NTNU, packaging), Erik Slinde (inventor and researcher), Elin S. Valle, Line Torsvik (Tine R&D) and Lars Edal (Gilde, sausage maker). All in all, a group with a broad set of competences was represented, but note that, again, the fish industry was not represented. The level of competence in the fish industry was not particularly highly regarded. Coming from a top management position to a chaotic and open-ended innovation project was a mixed experience for Rekdal:

> I had been managing director, and suddenly I became responsible for a project group. They were motivated and independent, and concerned with demonstrating their excellence. Hence, I didn't have the same kind of leader position as in a line organisation. And in the organisation outside the project we lacked support. It should probably have been organised independently of Tine. (Karl Inge Rekdal, Tine R&D)

Rekdal was thrown into a group in which several participants had been working on the invention for a while. He found himself lacking the formal authority of a line organisation and, in addition, feeling lack of support from the rest of the organisation. Communication problems soon appeared in the team, as one product developer described:

> You really had to be responsible for your part, no one else would take care of it, and you had to be demanding, as there were many strong people. I think we talked about the same things, but we couldn't understand each other. Karl Inge didn't fit the role of project manager. He was used to having his secretary around, writing all the letters and facilitating everything for him, and none of us wanted to be project secretary. (Janne Haugdal, Tine R&D)

The project participants had been recruited to the project on the basis of their professional experience and expertise, and were not very keen on taking on administrative tasks. Moreover, the project manager did not a represent the project on a superior level. It was Bente Mogård that participated on the Tine Biomarin board and therefore served as the messenger between the board and project team. The reason for this was related to history, with Mogård being an established board member of Tine Biomarin and active participant with an internationalisation responsibility in the Umi No Kami project, while Rekdal was totally new to the organisation of Tine Biomarin. There was a lack of common understanding, and problems of communication, involving much of the project team in a number of issues, especially when technological and marketing decisions were to be made. In addition, the grounding of the project within the greater organisation of Tine seems to have been weak, not in terms of expectations and attention, but rather in terms of top management commitment and capacity for following up. The team felt a lack of support:

> Business development and fish became just one of many projects. In my experience, there was no one in the corporate management who had real ownership to the fish ventures, and I think that is a weakness. There has to be someone committed to it, carrying it forwards. (Hilde Torvanger, Tine R&D/Marketing)

The project got resources, and they got a long-term perspective, but still they felt a bit isolated and left to themselves in the project, except when Rekdal or the CEO of Tine, Jan Ove Holmen, performed in media and other public venues, bringing with them the Umi No Kami project as their ambitious and promising new venture. After around one and half years, as a result of these problems, Rekdal accepted a CEO position in a company in the metal industry. Another turbulent phase followed, with at least two other persons testing the project manager chair, before Lars Petter Swensen was appointed to the job:

> I came into the project with the concrete task of developing methods for documenting raw material variations in the [fish salami] production. And then they needed a new project manager, and I knew the process so well, thus I went in as project manager. (Lars Petter Swensen, Tine R&D)

At the time Swensen was promoted, strong forces had started moving the project in a new direction (an issue that I will discuss below) and Swensen's new role played a crucial part in the change. It turned out to be more difficult than anticipated for the early team to draw synergies between the existing competences in the Tine system and new seafood ventures:

> We talked very nicely about these synergies, right, branding, our competence, our distribution channels, and even exports. But what we experienced along the way was that we didn't really manage to utilise, get full synergy out of it, because we placed the projects as satellites, outside the system. (Hilde Torvanger, Tine R&D/Marketing)

The problem was further described as a lack of ability to mobilise internal professional groups, whether marketing, management or technology, for helping out when needed in the project. Or as described by Haugdal as being a more general problem at Tine:

> It is difficult, you know, at Tine, that when you work on a project, at least in such an innovation project, then it is difficult to get other people to help you, who have the competence. (Janne Haugdal, Tine R&D)

This was experienced as a problem, not only between agricultural and biomarine activities, but even within the portfolio of fish projects:

> We had many different fish projects, and none of them were coordinated, we sat separately, and worked as best as we could. And, clearly, when I participated in the Umi No Kami project, we quickly discovered that we needed more competence on fish. (Hilde Torvanger, Tine R&D/Marketing)

Another option was to team up with external actors with knowledge of fermentation technology, fish processing and so forth; for example, when the sausage maker left the project, the product developers wanted to hire someone external, but the answer was negative. Haugdal and Torvanger described how this had consequences, possibly for the technology development time of the product and in uncertainty in moments of deciding on the further direction of technical issues. How ironic was it, then, to try and protect the patent application and other knowledge developed in the project by moving the experiments from the Norwegian Food Research Institute's lab at the university campus to Tine's own R&D premises:

> It was invested in a drying facility; we needed to be able to produce it ourselves. This indicates that we didn't trust doing these experiments at The Norwegian Food Research Institute, because when we now had gotten the rights to the product, we also had to guard the knowledge that came from the product experiments. (Per Magnus Mæhle, Tine)

To protect some kinds of knowledge, they dissociated other important knowledge from the project, even though this lack of trust concerned one of Tine's closest knowledge development partners for decades. On the industrial side, the search for potential partners had been going on since early in the project. Thus, they were left with a company-internal organisation of a relatively 'radical' innovation project, while the question remained about how to go ahead, when both the product and its market are unknown from the outset.

Yet, from there, how should one get people in marketing and management interested? On many occasions, I heard people from Tine R&D talk about their frustration about a lack of influence towards product development and marketing processes, resembling a 'technology push' paradigm (if my invention is good enough, it will find its way to the market by itself). According to the director of Tine Ingredients, Gunnar Hovland, technologists needed to learn to follow their ideas more aggressively through the system, not giving away the role of representing an idea before it has gained sufficient strength and understanding in the system. What the technologists had showed considerable commitment to, however, was the process of solving and stabilising the technological challenges of the innovation.

Recombining materials and stabilising fatty acids

Taking a highly uncertain and controversial innovation through numerous professional, social and organisational trials, from R&D through marketing to end consumers, is an exhaustive effort in itself. It is easy to forget about the actual object under development. What was the status of the object, the fermented fish sausage, during this early and explorative phase of the project? If we go to the 2003 status report, three important points were reported on the status of the object. First, successful test production at Tine R&D's own laboratory had finally been achieved. Second, this had enabled going further into searching for partners, preparing for large-scale production and market testing. Third, plans were made for further product development: finish the recipe, choose bacteria culture, develop near-infrared measurements of fat content and colour, get approval from authorities and establish procedures for analysis (Status report, Umi No Kami, 2003). After struggling with the technology for some time, succeeding with test production had been a stage gate for moving on to interaction with potential partners and customers. Here, I will go into a bit more detail about the technological challenges that the status report referred to, including stabilising the production and the product in a technical sense.

Making fish salami is a relatively complex process of combining and stabilising a set of biological substances. Biologist Even Manseth explained the main process in broad terms:

> The fish meat is ground and blended, mixing in the various milk extracts, or whey proteins, and the bacteria culture. Then the right

temperature is secured, and then it is put into sausage skins. When it is put into the skins, you put it into the oven that both smokes and dries it, and then the process, the bacteria starts producing lactic acid, right, and then the acidity increases, and then you conserve the product, and at the same time you dry the product, take out water, getting less water activity, and then smoke it. (Even Manseth, Tine R&D)

A process of three main parts: supplying and preparing the right raw materials, mixing the recipe and, finally, creating the best conditions possible for the micro-biological process within the product. Three micro-biological substances have to work well together: water, fat and proteins. And three biological processes had to be managed to make a fish salami: stabilisation of fatty acids (by adding proteins), fermentation (by adding bacteria cultures) and keeping away disturbing micro-organisms (by strict hygiene routines). I will below describe some of the challenges related to these three biotechnical processes.

There are two intertwined problems with fish fat when curing fish: it is fluid and it oxidises easily. It is a matter of stabilising materials in terms of a number of factors: texture, keeping them in place physically, maintaining nutritional and aesthetical value and helping them avoid becoming harsh. During the first attempts at making a fermented fish sausage, the fat did not stabilise and tended to slip out of the product. As mentioned previously, this was the reason why Erik Slinde asked Berit Nordvi and Tine R&D for help; he thought that adding proteins (e.g., from milk) could help solve the problem. After Tine bought the patent application and started the Umi No Kami product development project, the primary task was to take control of the technology. This quest for getting the technology to work led the technical project participants through a number of issues. In the research-based project prior to Umi No Kami, Neptun, they produced knowledge on the use of milk proteins to stabilise fatty acids. Slinde's hypothesis was right, proteins could be added in the processing of fish salami to stabilise the fatty acids. Still, there was a limit to how much fat could be stabilised in such a product, a limit that made it difficult to use only salmon (with its high and variable content of fat, 10–30 per cent) in the recipe. A number of different material and technological issues were investigated to enhance this stabilisation process. Different combinations of fat content and blends of saithe and salmon were tested, together with tests of other white fish species than saithe.

Transferring the technology and procedures from the practice of making salami out of meat, they used frozen raw materials to enhance the drying process.

> Frozen meat more easily lets go of water, hence one can take out the water faster. To work with frozen fish was simply related to the problem of anisakid nematodes.[9] Janne Haugdal wrote a letter to our Food Safety Authorities, to get permission to produce our product [from fresh fish]. They responded that farmed salmon was ok, as they have not found any anisakid nematodes on farmed salmon. (Berit Nordvi, Tine R&D)

In beginning with an unquestioned replication of the original procedure, using frozen raw materials, for reasons of drying, this practice was then reinforced by a problem of the new raw material, namely its common contamination of anisakid nematode larvae – which also demanded freezing before use. A lot of different species were tested in this version of the technology and recipe, without satisfying results:

> We have tested so many that I do not think I remember all of them. In the beginning, we started with saithe and salmon. And we did not get as fine of a colour. And therefore we tried to find other fish species that perhaps could bind the fat somewhat better, and get it stabilised so that we could get away from the big variance of fat in salmon, and at the same time it should be possible to use a white fish that gave a better colour and better durability after freezing. (Janne Haugdal, Tine R&D)

Even though the food regulation authorities would accept the use of fresh farmed salmon, white fish still had to be supplied from wild fish catch, and so the procedure of using frozen fish remained obligatory, if not going for the more difficult technological challenge of making a pure salmon product. At the time, the use of fresh fish was nothing but an interesting idea, something that, in the spirit of science, should be tested. Anyway, though whey protein proved its ability to stabilise some fatty acids, this process had to be improved quite a lot before having a satisfactory firm product. Thus, the testing of various white fish species continued, to see whether some combinations were more capable of binding fat than others, but the colour was not delicate enough. Then there was the problem of getting good enough quality on the white fish. The competing idea, of using only salmon – or at least as much salmon as possible – remained an option, and the food regulation authorities were asked for permission to use fresh (not frozen) farmed salmon. To enable management of the problem of fat content in a pure salmon product, near-infrared spectroscopy (NIR) was launched as a possible tool for helping choose the fish with the least content of fat:

> The challenge was to get a product that held together. We were allowed to start using this NIR-equipment, so that we could start sorting the salmon in different groups. Then we could use salmon with 10% fat, and get away from the large variance related to fat percentage, because you can't bind all the fat if the percentage reaches 20%. (Janne Haugdal, Tine R&D)

In this way, they could maximise the use of salmon in the recipe. Hence, Lars Petter Swensen, who had been working with this technology at the University of Life Sciences, was hired as consultant, and later became project manager for R&D in the project. Yet while technical success seemed to be moving closer, this did not help the economic aspect. The initial idea had been to use trimmings instead of premium fillets to keep down costs, and control of fat percentage in the fish improved from the near-infrared method (although this technology needed significant development before it could be used industrially), but still only the best parts of the fish could be used. This was also different from making salami out of meat, where trimmings are a common resource in the recipe. Haugdal summed up some of the similarities and differences between fish and meat related to fermentation:

> When you make cured sausage of meat, it is recommended that you freeze the meat, since you are supposed to destroy the proteins. You are not to bind that much water, as the water needs to get out of the sausage, since it is going to be dry. We froze the fish, but the product became too dry. But then we found that when we used fresh fish, it bound up more, and when we used pre-rigor fish, which is of the best quality with regard to proteins and binding capacity, then you got great sausage. The meat and the fish behave very differently. (Janne Haugdal, Tine R&D)

After two to three years of research, Haugdal could be quite specific about the transfer of meat technologies to fish. While the bacteria culture worked in the same way with fish, and the following pH and drying processes too, the process of binding fat was a lot more complex on fish. They had to strictly control the fat content, add proteins to encapsulate and stabilise the fatty acids, and use fresh premium raw materials instead of frozen trimmings. In total, these were significant changes for adapting the original technology to the new raw material, making the product both more expensive and more challenging to produce.

Still, if the use of frozen meat was such a taken-for-granted part of salami practices, and this matched well with the demand of freezing fish to avoid parasites, how did they come up with *the idea* of using fresh fish instead? This came from the growing group of aquacultural researchers at Tine, who were all recruited from a particular research group at the Norwegian University of Life Sciences (UMB). In addition to Swensen, Per Olav Skjervold had also been hired to a key position for the biomarine activities at Tine R&D, and when he called one of his junior researchers at the university, new ideas came up:

> When I did my PhD at Ås [UMB], Per Olav called me, and they had problems with some things. It was to get a good texture on the sausage, and they used, at the time, frozen raw material. I had read in the literature in relation to fresh products, that you have properties that you won't find in less fresh materials, so, I suggested more or less to try out fresher raw materials, closer to the farm, you could say. (Even Manseth, Tine R&D)

Manseth's rationale was very different; instead of starting out with the particulars of meat and fermentation technology, his perspective came from what he knew about fish, and how this perhaps could compensate for problems with the original technology. Hence, Lars Petter Swensen and Per Olav Skjervold started doing experiments with fresh salmon in addition to the 'official' experiments still using frozen materials. In Swensen's view, this change from using frozen fish (white and red) to fresh salmon was a seminal breakthrough of the project:

> The first quantum leap, as I see it, was related to getting control of the raw material. And the next quantum leap was the use of fresh fish, which became very interesting when we went from using 30-day-old fish, to using 24-hour-old fish and then to using 4-hour-old fish. We did a test, where I went over to Bremnes and got fish that was only six hours old, and we got a very compact sausage that did not fall apart, and that we could slice. (Lars Petter Swensen, Tine R&D)

Here, he was referring to using near-infrared technology to get control of the raw material, and then using the pre-rigor processed salmon from Bremnes Seashore, as two technical breakthroughs in the project. Using NIR to sort the raw material according to colour and fat content enabled maximum use of salmon in the recipe without including too much fat, and furthermore strengthened the visual presentation of the product with a stronger red colour. However, as Nordvi remarked, the NIR technology was at the time in a very early phase, with only a manual handheld instrument, not efficient, robust or reliable enough for putting it to large-scale industrial use. A few years later (2006), this technology had become available for automated industrial production, but still had not gained the argumentative strength – from an economic and use perspective – to support a purchase decision from the venture's owners. The second 'quantum leap' according to Swensen, was the breakthrough on what raw material produced the best results. Through the close relations formed through earlier projects to Bremnes Seashore and their pre-rigor processing technology (probably in having great interests in finding use for this technology), they went through a process of testing increasingly fresher fish. From these experiments, they ended up with dramatically improved texture of the product. Later, when scaling up production, moving the production from Tine's lab in Oslo to Bremnes Seashore's facilities by the sea, they were able to improve the freshness of the fish using another two hours – producing a product that is ready processed within four to six hours from slaughter. To Swensen, this represented 'paradigm shifts', moving the project to a 'totally new platform', from which they could start working in a more 'linear' product development fashion. No more need to 'mess about with other types of raw materials and other types of processes', thus leaving behind the original idea of using both white and red fish, and the practice from the meat industry of using frozen materials. Yet it also meant producing a product that was dependent on expensive and exclusive raw materials.

The technological basis started taking shape, not only for going into a more incremental product development phase, but also for moving the product from R&D to production. At the same time, the project also started its transition from the original project group to the hands of an emerging constellation between some university researchers, Tine management and Bremnes Seashore. This again stimulated the move from being a fermentation technology/fish salami project to becoming a fresh and cured pre-rigor salmon project. The project got both a 'rigid research design' and some very explorative aspects. It was a matter of simplification, keeping some variables stable while playing with others:

> Before we had great variation in the product, and because of these variations, it has been difficult to conclude on what results being the right, then, what development that has been right. (Lars Petter Swensen, Tine R&D)

The outcome, the 'quantum leaps', could be used for another set of powerful simplifications, including choosing to use only one type of raw material, from a supplier with close relations to the members of the project group and, in the long run, for simplifying the project intentions by choosing direction and developing a more unitary product and concept. Faced with all the material interactions in which a number of combinations proved difficult, impossible or ambiguous, finding a combination that worked had great consequences for the concept development. Swensen told about 'rigid design' and plans, but at the same time the project allowed, or required, lots of adaptation – or even transformation – of the technology and the concept.

Recruiting and controlling micro-organisms

Berit Nordvi, the project manager of Neptun, spoke about how they had worked on the fermentation technology, how the taste and texture of the fish salami were explored and successfully stabilised, and then forgotten and regarded as being unimportant by its later representatives. The more commonly told stories of the Umi No Kami project centred on the application of whey proteins for the stabilisation of fatty acids, and the choice of types of fish. Nordvi, however, emphasised the exploration of the original idea, and perhaps the most central technology in salami, namely fermentation – adding bacteria cultures in order to 'cure' the product.[10] Historically, the fermentation of foods such as meats has been a technique for enabling the product to be stored over periods of time (Ferrières, 2006),[11] and Tore Teigen, consulting sausage maker in the Umi No Kami project, could tell that at local pre-industrial farm dairies, bacteria cultures were hardly conceived of, or even known about, as they were stored and brought forwards through the un-/intentional storing of elements from previous productions: 'Before the 1970s, bacteria cultures were residing "in the house", or in the liquid solution from the previous production, it was not something they "made"' (Tore Teigen, sausage maker). Like 'magic', the transfer and maintenance of tribes of bacteria cultures happened through ritual-like procedures and practices – for example, adding some of the liquid solution from previous productions to the new ones, or sometimes the production room and facilities 'by themselves' served as a place of maintaining and transferring bacteria cultures to every new production. When industrialising production, however, the sterility of production facilities made this cultural element, the 'magic' effect, disappear. In addition subsequent scientific research within micro-biology has made it possible to work on cultivating some tribes of bacteria, while fighting others in enabling and controlling fermentation on an industrial scale.

Making fish more like a meat product, or employing a meat-curing technology in processing fish in an industrial company, demanded scientific exploration and testing, in this case study starting out with the core technology – fermentation. So, what about the bacteria cultures? What role did they play in this project, and who had the task of taming them and making them work? This time the bacteria did not work on their usual material, meat. Berit Nordvi related that, relatively early in the project, they had done a thorough laboratory testing of around 15 different bacteria cultures, mostly lactic acid cultures, based on the hypothesis that choice of bacteria culture could make a significant difference in quality, both with regard to the taste and stability of the fish product. According to her, they found one culture that produced clearly better results in both taste and texture than the others, an old culture that had been used with good results even in meat production. The later improvements of the recipe had some effect, but, in Nordvi's view, nothing close to the effect from choosing the right strain of bacteria:

> It was not as good as the products they have today, but it was still fairly good. And we did consumer research, a bit later, on salmon and saithe in combination, compared to salmon, showing that the difference was not that big. (Berit Nordvi, Tine R&D)

While the choice of fish species was not easily solved before meeting the Bremnes pre-rigor salmon, lab work here provided a clear positive answer. The chosen bacteria culture was immediately implemented in further work on the fish salami. The industrial logic of 'economies of scale' was activated, and testing was initiated to find whether the bacteria could be (re-)produced by Tine themselves, or if it would be better to outsource to large and specialised companies:

> We had people who worked with bacteria cultures elsewhere at Tine and who participated in this discussion, and tried to stipulate the

costs. It was calculated by the Stavanger-group, having some fermen-
tors, if they were able to do it, but they could not do it sterile, well, you
need specific equipment to do it this way. (Berit Nordvi, Tine R&D)

Just a few months earlier, Tine R&D had done a project to enable the
production of bacteria culture for the Jarlsberg cheese in-house. The
need to maintain the exclusive ownership to this culture had been
challenged by Tine's expansion of production to new plants in US and
in UK, and they needed to develop their ability to supply these factories
with ready-made bacteria cultures. It turned out that a machine, a 'filter fermentor', developed in yet another project, together with SINTEF,
a technology institute, proved useful to the task.[12] After some considerations of the cost and quality, they decided to try and do the production in-house, particularly to keep the ownership and knowledge of the
culture. By using existing knowledge and technology at Tine, developed for covering the rapidly increasing demand for Jarlsberg cheese
around the world, Tine R&D managed in a relatively short time to also
produce the culture for the fish salami at a large scale at their R&D
facilities. But still there were other micro-biological processes they had
to get control of.

In the minutes from a Umi No Kami project meeting (2003), it says
that 'experiment no.5 is soon ready, no mould problems, awaiting
results on visual, taste, quality'. Further, it says that personnel from
the marketing department, Frode Fimreite, would help in organising
market testing, something that had not been possible before because
they had not managed to produce any successful products. In the previously mentioned status report, the project manager could tell that
'product development [has been] delayed because of technical problems
(mould problems, 6 months during fall 2002)' (Status report, Umi No
Kami, 2003). Behind this written information, we can find one of the
most traumatic time periods in the project's lifetime. Six months of
hard and frustrating work of trying to avoid mould in the production;
identifying and removing sources of unwanted micro-biological activity. Six months' delay in the project, not managing one single successful
production, thus not being able to test the product in the market. The
Umi No Kami project group had early been sceptical of the limited time
frame for the project. Within one year, they were to both gain control of
the technology, and develop the product towards commercialisation:

It was the first thing we discussed in the first meeting we had, that
the time schedule was much too short. We would not be able to

launch anything at the given point of time. We got approval for a six month delay. Every batch takes a month to produce, which means that you can produce ten batches in a year, and then you have worked continuously all the time, and that is a lot, right. You probably won't manage ten batches either, because you then would have to work on the project all the time. (Janne Haugdal, Tine R&D)

Given the long period of production of each batch, the limited capacity for running parallel batches and the general organisation of R&D activities, with personnel often assigned to several projects at the same time, the team saw that this was a very tight deadline, and managed to negotiate another six months. On the other hand, they were also conscious of the significant economic investments that had been made to establish the project, so they felt pressure to produce results as fast as possible.

As mentioned previously, when establishing the Umi No Kami project, the lab experiments were moved from the Norwegian Food Research Institute to Tine's own lab at Kalbakken in Oslo. Nordvi emphasised the practical benefits instead of the aims to protect knowledge:

There were both economic reasons, because we expected a lot to happen in the project, and so we wanted to do it internally to protect knowledge, and also to be able to run production whenever we wanted. This is because there are often waiting lists at The Norwegian Food Research Institute. (Berit Nordvi, Tine R&D)

Thus, there were several benefits of doing everything in-house. First, there was increased flexibility from being able to produce whenever one wanted, independent of available capacity at the the Norwegian Food Research Institute labs. Second, economic arguments were mobilised, suggesting that in the long run, it would be cheaper and more efficient to be able to do the testing in-house. And, third, it was argued that it was improving the protection of new knowledge from the supposedly more open environment at the academic research institution. To enable this transfer, new drying facilities had to be bought and installed.

Still, the benefits of transferring lab experiments from the Norwegian Food Research Institute to Tine's facilities at Kalbakken in Oslo soon were overshadowed by technical problems. The supposedly rational act of changing labs, turned out to bring with it severe problems. Suddenly, they could not produce a single batch of fish salami without getting it covered by mould in the drying facility:

> We had a very difficult first phase. We had big problems with mould, and so we had to work a lot on procedures, both to keep the product clean, and to change things, in order to manage the product. And then it was not exactly an advantage having a bakery next door. It was a very tough start for those women, extremely frustrating for the project manager, and for all of us, there were frustrations and more frustrations. (Berit Nordvi, Tine R&D)

Those responsible for product development in that phase spoke of autumn 2002 as extremely frustrating, in repeatedly experiencing setbacks, and not managing to identify the reasons and fix the problem. A set of different explanations for the problems was launched by some of the participants. Erik Slinde pointed to the climate (drying) facility as the source of the problem:

> They did not know very much about climate facilities, because it was not the meat industry standing there, but the dairy industry, so the hygiene and everything like that was surely ok, but, there was fish hanging inside, and not milk. (Erik Slinde, Institute for Marine Research)

The project manager for the Neptun project, Berit Nordvi, agreed with Slinde about the significance of adjusting the drying facility, in addition to the issues of hygiene:

> So, really, that whole fall was about knocking their heads on the wall over those problems, and not getting any further. The first production that was ok, was in January 2003. Some of the reason were hygiene, and then some of the reason had to do with the drying facility, it was not adjusted optimally. And so all the products that I made at The Norwegian Food Research Institute are much better than what they could make at Kalbakken afterwards. (Berit Nordvi, Tine R&D)

Nordvi was here connecting the mould to the accumulation of water in the machine, which was something that had to be solved by adjustments. Janne Haugdal, product developer and one of the central participants in this phase, admitted having some problems with adjusting the drying facility, but nevertheless dismissed this as an explanation for the mould problems:

> No, it was the air and everything around it, and the equipment, we produce it cold, right, and then the sausage machine and the grinder have to be totally clean, everything has to be covered by foam, and you have to, for you kill nothing just by warming up the product like you do with normal sausage. (Janne Haugdal, Tine R&D)

In addition to pointing out these hygienic issues, being the product developer responsible for solving the mould problem, Haugdal wanted a professional sausage maker on the team:

> I am not a sausage maker or anything, but still I've got the responsibility for the product development part in Umi No Kami. It is important to have someone who can stand and feel the mince, and that knows cured sausage making in his hands. I know a lot about how ingredients are working, and I can even run a meat grinder, but I don't have it in my fingers. (Janne Haugdal, Tine R&D)

Haugdal's management thought this was an exaggerated demand. Haugdal, on the other hand, suggested that the experienced professional, knowing how a good salami mince should feel between his fingers, and how to put it through the various processing machines, could have expediated them to success. The embodied knowledge of an experienced craftsman was lacking. On the other hand, to what extent would such knowledge be relevant on a totally different raw material? Further, in her account of the mould problem, Haugdal emphasised air pollution (mould and yeast spores from the bakery next to the lab), hygiene and, to an extent, lack of experience with operating the machines as the main reasons for the problem:

> From the outset we had one year or something until we were to have a finished product, and when we struggled half a year with the mould, and could not manage a single test, then the frustrations grew bigger and bigger. And we disinfected all the equipment, and we bought a lamp to put into the drying facility, disinfecting inside the climate facility, and we made a whole lot of preconditions to enable a good product, then, and at last we made it, but it took us several months. (Janne Haugdal, Tine R&D)

Together, these factors formed a complex problem. It took them seven months to get there – from installing the drying facility to producing the first batch without any mould on, seven months delay before being able to continue in any way with the product development and the market testing work.

So, the mould was, in different versions, explained by the early project participants by hygiene and air pollution. Technological adjustments were admitted to represent a challenge, although deemed irrelevant by some of the 'insiders'. In addition, the quality of the fish was an issue, perhaps not for the development of mould, but definitely for making a sausage of a quality that could be eaten. The fish sector was often described as a 'cowboy' industry, with little competence on the most basic industrial issues, such as quality management, nutritional standards, knowledge-based product development, distribution and branding. One of the big problems was described as the micro-biological quality of fresh fish supply:

> We measured too high levels. We had to give them feedback that we couldn't use their fish. We controlled every single fish they sent us, and they were not good enough before we got fish from Bremnes. That was the first time we got proper fish quality and did not have to throw out a whole batch after production. (Janne Haugdal, Tine R&D)

Some of the participants in the Umi No Kami/Salma projects did not have much belief that it would be possible to solve this problem together with the fish industry. Haugdal, on the other hand, did not agree with this:

> By making strict demands on the suppliers of raw materials, I think that we would have been able to solve it. Just have to be tougher, can't just press on the quota, and you can't go via a wholesaler, you have to go straight to the supplier. (Janne Haugdal, Tine R&D)

Entering into a close and demanding relation with the supplier would possibly enable the quality Tine wanted, she argued. However, in practice, the desired quality was not achieved before switching to, and training, Bremnes Seashore as supplier.

Summing up: organising micro-actors

The wish to *internalise* test production at Tine, to protect and produce knowledge within the organisation, seems to have led to delays and

problems (there were no mould problems when running test production at the Norwegian Food Research Institute laboratory). They also
faced a loss of knowledge, when discontinuing the cooperative relations
with the Norwegian Food Research Institute/University of Biosciences
communities.

To sum up the main aspects of technology development, four intertwined technical and biological problems had to be solved before managing to produce the fish salami at Tine R&D's own facilities, with the
expected biological and nutritional quality and avoiding mould attacks.
Hygiene was the main and overarching issue at play. Routines had to
be sharpened, cleaning had to be done with extra care, and all of the
other issues were more or less related to these two. Second, there was a
bakery laboratory next door; hence, there were more spores from yeast
and mould in the air that caused trouble for the curing of fish. Third,
the new technology – the drying facility – had to be adjusted to work
optimally in relation to moisture. It was contested among the participants whether this had anything to do with the mould problems, but all
agreed nevertheless that it was something that needed to be done. Last,
the supply of – especially white – fish was a big problem. Several suppliers were tested, and they worked with some of them over time to make
them improve their micro-biological quality, without success. Until
testing the pre-rigor salmon from Bremnes Seashore. This issue was also
contested, as the Umi No Kami product developer clearly argued that
more work with, and stricter demands of, the white fish suppliers could
have produced good results, while others – particularly those who had
worked with Bremnes Seashore – were less optimistic with regard to the
competence and capabilities within the traditional fish industry.

In hindsight, three plain and simple things could have been done
differently, according to various participants, in order to simplify and
speed up problem solving during this frustrating stage. First, they could
have continued hiring a laboratory at the Norwegian Food Research
Institute where they started, and where the technological and situated
problems at Tine's own facilities did not exist. Still the supply problem
would have continued. A drawback would have been less flexibility, thus
possibly creating the need for more time than planned, though perhaps not as much extra time as the mould problem demanded. Another
potential problem was the protection of intellectual property, as the
environment of an academic research institute was viewed by some
as a potential threat for an expensive and prestigious project at Tine.
Second, it was suggested that they could have prioritised the recruitment of a sausage maker to the project team when they lost the one

they had in the beginning. A professional in the practice of handling minced meat, sausage machines and fermentation processes, could possibly have learned quicker about this new combination of fermentation and fish than the more general product developers in the project. And third, if Tine had been aware of the divergent nutritional standards and micro-biological competence in the fish industry, they might have evaluated the need for integration backwards with the suppliers differently: either by going to the actors with stricter demands and consultancy, or by formalising relations through acquisitions or strategic alliances, to get in control of the whole value chain from slaughter to final product; or by not doing fresh fish activities at all. In any case, Tine did buy the patent application, and they did have serious ambitions of commercialising the technology via a branded product concept. It is time to enquire into the customer side of the project.

Early market research and conceptualisation

From the early documents of the Umi No Kami project, we can learn some things about the original purposes of the project, and how the innovation was perceived by the main participants. Already in March 2001, almost a year before formalising the product development project (Umi No Kami), we can read from the Neptun project description how connotations to meat and markets for meat were made from the very beginning:

> It is possible to develop a range of new fermented seafood sausages,...that can be a real alternative to meat-based fermented sausages. The products may be optimised to satisfy different market demands...Use of the products: Similar to fermented products from meat. Perfect as snack together with beer, etc. (Project description, Neptun, 2001)

However, in the same document, starting to develop arguments for the consumer value of such a product (as opposed to the value for the researchers, which is more related to fascination and interests in managing a new technological combination) associations to the fish side were emphasised:

Food with positive health effects:

- High content of omega 3 (e.g., consumption of 40–50 gram fish sausage (contains 20% fish fat) covers recommended daily portion.

- High content of A and D vitamins.
- High level of positive lactic acid bacteria cultures.
- Fish (and not disease-infected meat).

(Project description, Neptun, 2001)

Associations were made to health trends, the recent scientific research on the benefits of marine omega 3, and the quest for clean, uncontaminated foods: the market for fermented meat, combined with the nutritional value of fish. To get off to a good start, the project team started developing a business plan for the product right after buying the patent application and formalising the project. Formally, the object at the time was little more than a patent application, but, in practice, Berit Nordvi and others had been able show it to people, taste it and experiment with various ingredients. Thus, at the time, there was a relatively close association between the patent application's prescriptions for fermenting fish, and the temporary status of the object, experimenting with materials and technologies in practice:

> The product was then not a pure salmon product, because the whole patent idea was based on getting enough proteins, and so it was a mix of saithe and salmon. What was presented was one to one of this, and it was used milk proteins, whey proteins. It was coloured, and looked nice. (Berit Nordvi, Tine R&D)

Although they made a few adjustments, like adding colour, they followed the recipe of the patent application closely when making the material prototype. The first 'strategy document' sought ways to associate this emergent object with the industrial reality of food – that is, established categorisations of food items, potential consumer 'needs' and the international structure of food distribution:

> We can develop a fish product that is somewhat similar to fermented sausage in its texture, taste, and shelf-life. This type of product does not exist in the market today. Thus, nor does any natural category exist in the food trade where this product belongs. Our challenge is threefold: Develop a concept that is sustainable over time. Develop a product and a product series based on needs in the market. Find those markets with the highest potential for this type of product. The concept will be the start of an international brand within fish products. ('Strategy document 1', Umi No Kami, 2002)

Here, we see that the identity of the original invention was a 'fermented fish product', 'similar to fermented sausage', in other words, a salami of fish. Yet the market category for such a product was at best ambiguous, as there was no 'natural category' to which this product belonged. It could be a fish product, or a fermented meat product. A little later, the ambiguity expanded to include consumers and their potential *use* of such a product: gourmet 'upper class', or mundane everyday product? The identified challenge was, however, as clear as it was difficult: develop concept, product and market at the same time. A SWOT analysis followed in the document, emphasising, on the 'strengths' side the patent, the healthy aspects, the long shelf life and the need for cold-chain technology, which is considered to be a core competence at Tine. On the 'weaknesses' side, the uncertainty of what consumers wanted, its potential for use in warm dishes, the lack of a product category and the problem of identity – the fact that a 'mixed' product processed into a sausage did not communicate freshness or highest quality – were mentioned. The opportunities that were mentioned were the possibility of establishing a new category and fitting the concept within various food trends (Thai, Italian, etc.). The most threatening issue was viewed as being the problem of categorisation, as the product would either drown in the amount of new products, or the customers would not understand what it was or how it could be used. Due to considerations of risk, it was suggested that 'Tine' was not used as main brand, but rather either a 'supported or pluralistic (independent) strategy' was chosen.

With the formalisation of the Umi No Kami project, it got fresh resources for exploring and sketching a product concept by seeking knowledge of consumers and their reactions to this product. The first sketch of a marketing plan reflected on why people would purchase and re-purchase the product. 'Curiosity and health' was mentioned as the main triggers for customers, again linking Umi No Kami's content of omega 3 to the growing health trends within food ('Strategy document 2', UNK, 4 June 2002). However, at this point the marketing plan contained no references to concrete market analyses, and looked more like an expression of the project group's own speculations and reflections. Two months after the first strategy note, a business plan was ready, outlining in its introduction a future scenario for the use and diffusion of utopian dimensions:

It is strange to think back on the starting point for Umi No Kami. We had purchased a patent application. That was all. But it was not any patent application; it was a real innovation, and not just an idea for product development. We knew that if we could manage the idea right, we would be able to launch a unique product concept. Not just in Norway, but also internationally. And today, 18 years later, we all know what success this has become. It is a well-known product to be found on all breakfast tables and in all sandwich outlets. When we look back, it is almost like the product has always been there. It has grown into our diet as a missed element, and has been adopted and accepted as part of our food tradition. (KIR and HT 2020) (From 'Business plan', Umi No Kami, 2002–7)

We can almost feel the optimism and fascination the project team must have felt for the innovation. In this document, a mass market is projected, imagining how Umi No Kami would become an everyday product 'on all breakfast tables and in all sandwich outlets', and the rest of the business plan is made to support this future scenario, from market estimations and considerations in budgeting. No associations were made to appetisers in restaurants, or for special occasions in homes with a high level of education, like the first strategy document written two months earlier. Further, there were no traces of the ambiguity of market segments, product identity and technology evident in the earlier notes, which followed the project throughout its lifetime. A year later, in a status report given by the project manager, the potential *use* of the product was downplayed, instead focusing on the geographical aspect:

We think we have to go for a core product, and then adapt some products and variants to local conditions. . . . The purpose is to develop a series of fish products that take part in creating a whole new category of fermented and dried fish products in the food trade, that are profitable, and that the consumer wants. The product is to be sold both in Norway and internationally. (Status report, Umi No Kami, 20 June 2003, by Karl Inge Rekdal, Tine R&D)

This is clearly intended to be an international project, as this report does not show anything other than associations to fish, and maintains awareness about the challenge of establishing a new product category. In the first quote, a tension was identified between a 'core product' – that is, the need for simplification of the product development and production processes (economy of scale) and the need for local adaptations to different geographical markets and market segments. There was an awareness arising about this strategic question, which was later transformed into a set of socio-material strategies of product design, production facilities and sales activities. The second quote shows how the project group had been working, not just on the problem of categorisation, but also on the discursive associations to the emerging object. They started creating its context, so to speak, by outlining a portfolio of product variants that together were meant to produce a 'whole new category of fermented and dried fish products in the food trade'. Moreover, it would become profitable and something 'the consumer wants'. This statement points both towards the need for creating customer value in order to succeed, and towards Tine's (and Tine R&D in particular) parallel work on reconstructing their conception of their own general knowledge – that is, not to 'process milk', but to 'pick the raw material apart, and then put it back together in the way that the customer wants it' (Per Olav Skjervold, Tine R&D). In other words, they were creating meaning in the project by connecting it to the corporate strategy process.

A strong line of thought in the early argumentation for the project was how to secure internal support and resource access. This 'internal marketing' work can be traced as being implicit in various sources (some of which have been alluded to above), but there are also quite explicit accounts of this aspect of the innovation process. This is not to say, however, that the participants did not believe in their own argumentation:

> And when I look back on this now, I am almost embarrassed about what economic expectations we had to this. But I was very excited, and had incredible belief in it, and I still do, I am certain that even the Norwegian market has potential. (Hilde Torvanger, Tine R&D/ marketing)

Although acknowledging a few years later, after having left the project group, that they had been too optimistic, Torvanger was still excited about the innovation and had hopes for its success. Yet when confronted with the above-mentioned 'hairy' goals in the early strategy notes, she could explain more about the dilemma of early-phase innovation:

> Clearly, the dilemma when it comes to big investments, right, is costs, then you have to show expectations of great profit. But the question is, when you pursue a radical innovation, like this, what kind of expectations should you have? Should you think from a seven-year perspective, or from a three-year perspective, as the management at the time demanded? And in my opinion, it is totally unrealistic to think from a three-year perspective, as it would be too little time for

getting profits after 3 years. It wasn't any problem for me, because I believed in it. (Hilde Torvanger, Tine R&D/marketing)

They knew the project would take time and money, and, to legitimate this use of resources, expectations of great profit had to be demonstrated. Stories of future prosperity had to be put forward, which were partially grounded in the product and their limited knowledge of its potential markets, but mainly based on the expected level of ambition within the company. If three years are what a company is allotted, then three years are what it starts out with, in the hope that it will be sufficient, or at least that management would gain enough commitment to the project by then to continue supporting it. Berit Nordvi could confirm how they worked to sell in things to Tine:

The main aim with that pre-project group, which I participated in, was to sell in things to Tine. This was a project that was discussed back and forth, and there was a lot of uncertainty, so one of the ways we handled it – at Tine's national meeting – was to serve the product to them, and they ate. In addition, we ran a consumer test, with focus groups, and got an evaluation of what people in Norway thought about it. The second step was to do a market survey by sending out the product around in Norway, early spring, to get feedback on the product. (Berit Nordvi, Tine R&D)

They acknowledged that a fish salami would have difficulties finding much enthusiasm, or even interest, in the Tine organisation without a demonstration of it – letting people taste, smell and feel it, use their senses to evaluate what their sceptical brains otherwise would have rejected as nonsense. Second, they sought to back up the physical experience of the product with argumentation based on consumer data. Thus, focus groups and market surveys (i.e., home testing of the product) were run to produce evidence of market potential.

Types of customers and quality/price

Who would want to buy salami made of fish? The great interest and enthusiasm in the dairy cooperative for this hybrid object was derived for the most part from the technical and biological problems it promised to overcome:

And to bind the fatty acids in fish was very exciting, but this is technology driven, right, because to get hold of, or to master something new in this way I think triggered a lot, but at the same time we saw that there had to be some opportunities in the market for this. (Hilde Torvanger, Tine R&D/marketing)

'Mastering something new' triggered motivation, which again triggered a search for arguments in favour of the project, as 'there has to be some opportunities in the market'. Yet despite having been bought by a commercially oriented corporation, this project represented nothing but uncertainty when it came to potential customers:

But it was a recipe for something we didn't know what would become. It was called fish salami, but what on earth could it be used for? We didn't know, but we just saw that it could mean many opportunities, a great potential. (Hilde Torvanger, Tine R&D/marketing)

When they could find ways of solving these biological problems,[13] then – surely – someone would be interested in buying the resulting products?

I think that the sandwich filling was the first priority, that we at last could get a sandwich filling of fish to the people, in a way, because that is almost non-existent today, you have pâtés and smoked salmon, but smoked salmon is not an everyday product, right. (Hilde Torvanger, Tine R&D/marketing)

'Sandwich filling of fish for the people' fits well within Tine's portfolio of dairy products which have been on every Norwegian breakfast table for decades, an 'everyday product'. Yet, did it fit in with these users' perceptions and use-practices? Two of the main supporters of Umi No Kami (throughout the project) within the corporate management were confronted with the issue of involving users:

The problem is that where you have genuinely new products, and these are products that the customers have substitutes for today, then the customer finds the product interesting, but will the consumers be interested? Is this something that will sell? So, we have tried, to a large extent, to get customers in, and have them take a role. But it is pretty hard on these kinds of products. (Per Magnus Mæhle, Tine)

The way of conducting user-involvement in this more radical project was not routine, even for a company with 100 years of experience with incremental development of dairy products.[14] Thus, they both lacked established methods and practices for involving customers in radical innovation, and knowledge about how to handle the ambiguity of customers related to the initial interest and final purchase of new products. Note also the distinction made here between customers and consumers. While business customers (retail, restaurant, etc.) showed some initial interest, they were still not easily involved in the project. Further, while business customers expressed some interest, it did not mean that the end-users – the consumers – would be convinced to buy anything. Refsholt further problematised the issue of when and how to involve customers in the development process:

> How much R&D work needs to be done before you have something to show the customers? So that you can establish customer involvement, and how significant is the development work during customer involvement? I think we have an increasing awareness of the importance of doing this early. But it is very difficult to do before you have something. You need to have something concrete. (Hanne Refsholt, Tine)

While acknowledging the need for early involvement and customers, Refsholt nevertheless questioned the timing of this. It was seen as difficult to involve external actors before having something concrete to show them; a prototype, or a demonstration of possible versions. An abstract idea was not enough. For the project group, this dilemma ended in a rather defensive strategy related to industrial customers (e.g., retail): We had not come far enough to dare to contact the retail chains, and this had something to do with confidentiality too, right. (Hilde Torvanger, Tine R&D/marketing) 'To dare to contact' retail chains both had to do with how to present something so unfinished without making a fool of itself, and with confidentiality. The fact that this was brought into Tine via a patent application, made them conscious of IPR issues and it sought to keep the knowledge developed a secret until approaching the time for launching it in the market. Yet there was one group of users who had been greatly involved during most of the project, namely chefs:

> And we worked with chefs, we had several sessions with the Culinary Institute, and with the chef at Neptun, he is experimental. So we obviously gathered a great deal of food competence, in relation to how you can use the product, and what it is suitable for. (Hilde Torvanger, Tine R&D/marketing)

A chef with experience from the Culinary Institute and various restaurants, Svein Erik Hilsen, working at Tine R&D, was participating on and off on the project. He was both an important advisor to the project group and a bridge to other experienced chefs. This was instrumental in evaluating recipes and in producing dishes with the fish salami as a main ingredient for pictures and demonstrations.

Management preconditions and initial intentions

The management commitment to the project, or deciding to buy the patent and establish a product development project, did not mean that everything was then settled with corporate management. What were important for the direction of the project were its initial intentions, forming what we could call 'framework conditions', stemming partly from management discussions and decisions on buying and formalising the project, and partly from the patent application itself – its recipe and prescriptions for making this 'fermented fish product'. Both the technical side (raw material, processing technology) and the market side (mixed product, domestic and international) seems to some extent to have been conditioned by these initial intentions, limiting and shaping the potential pathways ('opportunities') explored by the project team:

> When we started this we had to use white fish, in a mix with red fish. This was decided as a premise, and then we had to stick to it. So, when we started developing a communication platform and name, we talked a lot about 'Sea Salami', and all such 'Salmon'-things, right, but this was out of the question, as we would then be limited exclusively to salmon. And in addition, this product was supposed to be international, and a gourmet product, right, we wanted it to be upscale, so we ended with a product name called 'Deli-Fjord', delicacies from the Norwegian fjords, right. (Hilde Torvanger, Tine R&D/ marketing)

Although she had expressed sympathy with these framework conditions when discussing the marketing strategy, Torvanger described how this also limited their work. The obligation to include white fish led to technical difficulties in stabilising the product technically due to low quality in raw white fish materials; its greyish colour was a challenge to associate with a 'gourmet product'. Finally, in the conceptualisation and naming process, they could not take into consideration things that hinted too much towards salmon. When questioned for more detail on these original preconditions and how they changed, Torvanger brought together several of the issues mentioned in this chapter:

> Clearly it created a lot of frustration along the way. There were enormous challenges prior, because they demanded [the mixed version], and this was very difficult. We were obliged to launch in Norway first, before going international. So, we worked hard to understand how this product, which originally was to be an upper-class gourmet product and which we don't have large segments for in Norway, how could we manage to create a success story in the Norwegian market that would be the basis for going abroad? We worked a lot with the sandwich filling market, to see how we could get it in there. You know, how can you change those two or three products that you always have in your fridge? You will probably have many other things once in a while, but you have just a few permanent ones. (Hilde Torvanger, Tine R&D/marketing)

In addition to the technical and conceptual challenges of mixing white and red fish, the obligation to launch in Norway first was here seen as problematic, in particular because an 'upper-class gourmet product' would not find enough customers in this small and price-sensitive country; hence, the production of a 'success story' legitimising going abroad would be challenging. Torvanger described the typical Norwegian consumer as extremely loyal to a small set of sandwich fillings, in a very competitive retail market. These associations to the mundane, everyday eating habits of Norwegian households were not easy to combine with an 'upper-class gourmet product', which was mentioned in that same passage. So, at the time, these preconditions were experienced as being very stable and restrictive, while later in the process, for some specific reasons, they could suddenly be changed. Yet how can you investigate the market, when being stuck with this obligatory framework condition at this stage, but still no finished product?

International study tour

> We worked a lot with studying what markets that could be relevant, related to the taste, right, it was a dried, fermented, and smoked fish taste, where would that be accepted? And where do they eat much fish, in other words, where could we get it out to many different segments? (Hilde Torvanger, Tine R&D/marketing)

Internationally, the Umi No Kami project group chose to map the international cultural geography of food – identifying areas with significant 'fish cultures' and 'cured meat and salami cultures', and then exploring these areas both by learning about their habits, preferences and distribution practices, and by giving out tasters of a few potential versions of their fish salami. The transformation of research interests into commercial products was, in this case study, obviously not a straightforward and simple exercise. This was therefore given some priority, in addition to the central task of getting the technology to work industrially. Responsibilities were given to project participants Hilde Torvanger for marketing, and Bente Mogård for internationalisation, and they organised a few study tours, to learn about various markets and get feedback on the fish salami from local 'experts':

> We were allowed several study tours, and studied Italy related to the Mediterranean countries, and we went to Japan and South Korea, which are also important fish markets. Clearly, I have been able to explain better afterwards. We had room for playing around with things. (Hilde Torvanger, Tine R&D/marketing)

Geographically and culturally, potential users and their contexts were investigated, with facilitation and help from Innovation Norway (the Norwegian Export Council). Under the commercial uncertainty of the project at the time, both the project team themselves and their fellows in Innovation Norway really felt that 'going to the market first' was 'a right way of doing things'. This view was further confirmed by the market actors visited:

> We talked to chefs in several different hotels, we talked to purchasing managers in big food stores, we talked to food journalists, and to analysis agencies, so there were a wide range of different inputs. There and then, the response on the product was very positive. But we got comments that there could be changes on this and that, that there are opportunities here and there, etc. (Hilde Torvanger, Tine R&D/marketing)

Both in relation to the new marketing dogma of 'customer-driven' innovation, and to the experienced need for learning more about potential product/market interfaces, this act of asking actors closer to potential markets was experienced as being useful. Various 'opportunities' and 'opportunity gaps' were found, enabling Torvanger to 'explain better afterwards' what the project was about, but on the other hand also maintaining the multiplicity, ambiguity and lack of focus by finding 'too many fields of opportunity'. Most project participants supported the idea of going to 'the market' before finishing the product:

> The taste at the time was not as it is now, so in a way we presented something we did not yet have, thus we kept the presentation low profile, and talked more about the concept. We got positive responses during these early market encounters. But, we understood that we should absolutely not call it a 'sausage', it would then be associated with a cheap mixed product. (Karl Inge Rekdal, Tine R&D)

To present something that one does 'not yet have' is not an easy thing to do; hence, Rekdal preferred keeping a low profile on the product. He nevertheless found the tours useful, helping him to make sense of what to (not) call the product, and how to present it. Different versions of the product were brought on the tours, to get broader input for choosing direction in the project:

> We brought two, three or four different recipes. We had one with a very high content of salmon. It looked much better, right, but the others had good taste too, but what looked good, you know, the others had a too grey colour. And at the same time we got insight into distribution systems, retail chain structures, shopping patterns, we got plenty of input. (Hilde Torvanger, Tine R&D/marketing)

Thus, the visual performance of the product made a difference to these market representatives. The greyish colour of the versions with more white fish did not make as good of an impression as the light red colour of those containing more salmon – even though 'the tastes were not very different'. The other learning output from these tours, was insight into the different 'distribution systems' – that is, how the fish/food industries were organised in the different regions – making it easier to be more goal oriented in further marketing work. However, when the time came to employ more common market research tools, they chose to focus on the domestic markets.

Domestic market research

While knowing the food retail structure in Norway through and through, the Umi No Kami project group knew less about how the domestic consumers would judge their innovation, thus mobilising Tine's apparatus for market research. After some frustrating months of work in the lab (see above), versions of the product were ready for testing by a number of Norwegian test users, to explore how a fish salami went together with Norwegian practices of cooking and eating, and how it fitted in refrigerators, in different meals, and among consumer preferences. However, first they chose to get the technology and recipe to work properly before doing anything more with potential markets and market representatives:

> To begin with there were a lot of fragmented market tests, and intense work in R&D. Then we said that 'let's cut all the other things, now we just have to work on making a good product, and when that is in place, then we can start looking at what to do'. And after around six months only focusing on the taste, then we had a good product, and we asked 'what can we do with this?', it surely could be used for something. (Gunnar Hovland, Tine Ingredients)

At that point in time, the director for Tine Biomarin, Hovland, had become impatient with progress in the Umi No Kami project. First, they had to make the technology work, which meant another six months of lab work before finally being able to control the crucial aspects of the production process, hence having *something* – a stable product – to present to the market. Then they could start doing more systematic market research, such as focus groups:

> And we worked with focus groups, and we really tried to go in and see how consumers would use it, tested it, simply served various dishes, and got them to taste and use it in different ways, and see what was suitable for tapas, on sandwich, as ingredient in hot dishes, on pizza, so we tried many different ways. (Hilde Torvanger, Tine R&D/marketing)

And then, home testing in 300 Norwegian households:

> When certain improvements had been made, so that we had a product, it was sent to members of the Norwegian population. This was to test the potential for re-purchase, and they looked at the Norwegian and the Scandinavian market. [The report] said that it would be somewhere between 'brown cheese' and mackerel in tomato sauce, which is very large, but it is very difficult to make estimates on the potential volumes of a product. (Lars Petter Swensen, Tine R&D)

The results were relatively positive. Many of the consumers in the tests approved the product, and suggested how they preferred using it, while others found it too strange to integrate it in their cooking and eating practices. If these market estimates had been a correct forecast, this would have made the fish salami one of the most sold sandwich fillings in the Norwegian market, which would thereby make the project profitable before even trying to access international markets. This is all very well with the benefit of hindsight, but such assumptions cannot be taken for granted at the time, despite getting good feedback from the test users:

> They gave a pretty good re-purchase response, in the Norwegian market. So, this is the only, concrete, market analysis that has been made. But we know that, especially on new products, such market analyses do not have very high value. (Lars Petter Swensen, Tine R&D)

As Swensen indicated, market research during this phase of the project had limited value. It did provide some responses that the participants found to be useful; they also found it useful to learn about different regional food traditions in order to adapt the product to different markets. Further, perhaps most importantly, it did some work in producing 'evidence' for market potential in the internal marketing of the project, to Tine's management, owners, and other involved actors.

Summing up market opportunities

Inspired and informed by these events, the team identified some marketing opportunities they considered to have potential:

> Firstly, we were able to identify these different segments; the sandwich area in Europe, the snack and gift market in the East, and so on, and then we could narrow down, and then we could start putting numbers on it, how big is this really? We could buy numbers related to what we were talking about. (Hilde Torvanger, Tine R&D/marketing)

So, according to Torvanger, the research enabled them to start narrowing down and choosing direction. In summing up the project, Bente Mogård, who was responsible for internationalisation, displayed some of the reasoning during the process. The early market studies identified 'the areas in which such a product could work', particularly the fermentation traditions in Spain, Italy, Germany and the fish traditions in Asia, especially Japan and Korea. These were also identified as markets with buying power. However, then 'South-Europe was early put aside, as Norwegian fermented sausages not are comparable to the Spanish'. In realising how the success of new fermented products in this region would probably be closely connected to very local and culturally significant actors, they chose to go for Nordic markets on the European continent instead, in addition to Asian markets. Mogård not only saw the project as a single project on fish, but emphasised how they also 'used the project to do something with the internationalisation at Tine'. The Umi No Kami project was described as a learning arena for Tine's international activities in general.

Torvanger made some clear distinctions, mainly related to cultural/geographical variations:

> We had a focus on the east. A sandwich filling in Norway, snack in Japan, and the gift market is enormous, both in Korea and Japan. And in the Mediterranean and European area we had the fast growing sandwich culture, with lack of fish based fillings, a lot of tuna, but not anything else, and this was a clean and dry fish product. Ordinary smoked salmon is used, but is greasier, while this product was easier to handle. (Hilde Torvanger, Tine R&D/marketing)

The 'salami culture' of the Mediterranean had been ruled out, and so they were left with Umi No Kami as a sandwich filling for the Norwegian and European markets, as well as a snack and gift for the Japanese and Asian markets. The practical characteristics of being 'clean and dry' instead of the more 'greasy' smoked salmon and tuna had become a selling point for Europe. In Asia, Umi No Kami sought to be associated with the practices of eating more fish, and of giving red-coloured gifts. Some choices were made, but Umi No Kami was still a multiple – and hence ambiguous – object, both in relation to market segments and to product features. This ambiguity cannot be explained simply by these sets of 'market opportunities', as each of them would need strong representation by participants within the project to survive. It was, in other words, a negotiated aspect of the identity of the innovation.

Summing up identity negotiations

There were three different, although interrelated, negotiations about the identity of Umi No Kami. One was related to the various interests of the involved participants working on this fish salami invention:

> You had Neptun, which was a very research based interest, not product focused at all. And this was a very strong driver because

the researchers knew where they wanted, and had parallel activities aiming towards a PhD, and the other [product development] was a direction that did not know what it wanted. And then you had marketing, which neither had the understanding of which way to go. (Lars Petter Swensen, Tine R&D)

In the Umi No Kami project, the people coming from the Neptun project had clear interests in science and technology development, seeing Umi No Kami as a natural evolution of their research, even if they were aiming for commercialisation. Then there were the product development people working in a middle position between research and commerce, seeking to put together and stabilise a product that would work in industrial production and would hopefully find commercial use. The third group was those working with marketing, which had responsibility for researching potential markets, and beginning to develop a market concept for the innovation. According to Swensen, only the Neptun people at the time had a clear conception of what the Umi No Kami project should be about, and therefore more influence than the groups which had to struggle with the huge uncertainties regarding how and where to commercialise the innovation. It was brought into Tine in the first place by scientists interested in the technological matters of fermentation, proteins and fat stabilisation, and so these research-based interests retained strong representation in the project throughout much of the Umi No Kami project.

A second negotiation was between human and non-human elements; between researchers and product developers and their raw materials – biological substances of various kinds – in the battle for making the resulting product materially and technologically stable. The various fish species had different characteristics, or we could say preferences, which to various degrees were combinable with the other raw materials or with conceptions of how the product should look, smell and taste. The various micro-organisms at work also had their own interests; mould found the product extremely attractive and was hard to keep away, while the best tribe of lactic acid bacteria needed to be identified and recruited. As was described in relation to technology development earlier in this chapter, some things were rendered impossible – technologically not feasible – while other things worked out through adaptation work. This clearly influenced the conceptualisation work. Going for an everyday, low-price version of the product became increasingly difficult in the face of the increasing costs of high-quality materials and production technologies. On the other hand, the use of only salmon, and the qualities of the pre-rigor raw material, opened up new opportunities on the high-end side of the scale.

The third negotiation was related to the preconditions and directives from top management, shaping the perceptions of what the project group could and could not do with the project. I have earlier described the initial intentions of the Neptun and Umi No Kami projects, and how the patent application and top management decisions became a quite rigid frame for Umi No Kami. The project team reported having problems getting approval for changes in this framework, and also struggled to get extra resources. Then, after two years of technical and conceptual exploration of fermenting fish, a chain of events occurred that totally changed the situation. Suddenly, these preconditions could be renegotiated and changed. One version of the story was that it was triggered by changes in Tine's corporate management:

> But what happened in the corporate management was that the strongest advocates for preconditions such as doing this in Norway, that it had to be a sandwich filling, and those things, quit and went out. And then the preconditions fell too. (Hilde Torvanger, Tine R&D/marketing)

The corporate director for Tine's Northern region, Kåre Magnussen, together with Per Magnus Mæhle (director for R&D, and later corporate director for strategy/business development) and Ingrid Svensen (international director), had responsibility for the biomarine activities at Tine. When Magnussen quit his job, supposedly the strongest proponent for the initial preconditions of the project, suddenly the whole project could be re-evaluated. From working hard both for the domestic and the international markets, Norway was now removed as 'obligatory passage point' for Umi No Kami's marketing strategy. Furthermore, the transfer of the project from being a Tine R&D project to becoming a commercial venture within Tine Biomarin (which slightly later merged with Tine Ingredients) was a strong signal of change of focus – from product development to commercialisation. This transition became the start of a radical transformation of the project.

Reorganising the project and scaling up production

Along with the work on technical and conceptual development, work was also done on identifying and evaluating alternative ways of taking the Umi No Kami from the lab to large-scale production and distribution. Doing everything within Tine was considered to be an alternative, but their lack of knowledge, both on the meat and fish side, was viewed as being a reason for also considering partnering opportunities. During this process of partner search, a number of alternative solutions had been considered. Early on, the Norwegian Meat Cooperative, Gilde, had been consulted to check on their interests in collaboration. Gilde both had expertise on process technology of fermenting meat, and a dominant market position on cured meat products in Norway, but Gilde rejected the proposal:

> Tine has no experience with climatising. And we had a feeling that Gilde did not want competition, or perhaps we were not good enough in marketing the project to them. Gunnar [Hovland] and I could not reach them. (Karl Inge Rekdal, Tine R&D)

Gilde did not seem very interested in backing up commercialisation of the 'fish salami', a potential challenger of their core business. Another alternative for Tine was to buy a redundant meat facility, with the project manager as a strong proponent of this idea. Haugdal strongly opposed the buying of large-scale production facilities – former meat facilities – before succeeding with production in the lab. Those who were in the middle of the demanding and frustrating process did not find it wise to decide on industrialisation issues, like how to scale up production, before having full control of the basic technology. The product developers, still struggling hard to stabilise the technology, strongly opposed the project manager, and wrote a separate evaluation on the matter and sent it directly to the top management:

> I feel that it is my fault that we did not do this. It was a fine facility, and everything was ok, and they were supposed to close down their production. But it was all too early, because at that time we had all the trouble with the mould, and we had not yet managed to produce anything in the lab. I made a note on the pros and cons, and then it was decided to not buy it. (Janne Haugdal, Tine R&D)

These concerns about the unstable state of the technology in the lab, and the following uncertainty about requirements for scaling up the production, made the management decide to wait with further investments until the object was stable in small-scale production in the lab, in spite of the project manager's suggestion. Two other companies that were considered were Sunnfisk and Lofotprodukter. While Sunnfisk was disregarded relatively early, Lofotprodukter, partly owned by Tine, and already producing a number of different mince-based fish products, came to be seen as the most relevant partner:

> We evaluated two or three other companies too, amongst these were Lofotprodukter, and we had arguments on things like equipment, localities and facilities. Had they been supreme, then we would have collaborated with them. But when the raw material became the pivotal element for collaboration, then we made an objective evaluation of the other partners. (Lars Petter Swensen, Tine R&D)

This 'raw material' referred to the use of fresh pre-rigor salmon from Bremnes Seashore. From then on, access to high-quality raw materials quickly became the most important criteria, not only for making and explaining the product, but also for evaluating potential partners: 'When we started to get control of the raw material, and we started with fresh fish, it was Bremnes that clearly stood out as an obvious partner' (Lars Petter Swensen, Tine R&D).

With raw material quality as pivotal criteria, and with Bremnes' pre-rigor salmon as the benchmark, the matter became increasingly clear to the organisation. Thus, Lofotprodukter and other alternatives were turned down in favour of the provider of the best raw materials, and the project participants were – again – thrilled by access to novel technology. This choice and the process that preceded it also had some relational components. The newly hired university researchers had already a long-standing research relationship with Bremnes Seashore, and Nordvi felt that this influenced the choice:

> Obviously, it was based on close relationships. It was also an evaluation of Lofotprodukter, and of the facility outside Bergen. In fact, we sent our quality control people to evaluate those facilities, but in a way, they felt that they were overruled, and that Bremnes should be chosen. (Berit Nordvi, Tine R&D)

This view was confirmed by a member of the new group of researchers: '[We] have had a research relationship with [Bremnes Seashore]. This is probably the lever that has made Salma fresh loin, and all that, come about' (Lars Petter Swensen, Tine R&D).

Although being framed by the new group as an objective and open search for the best alternative, we also see how various interests and relationships were in part shaping the process. Whether it involved frustrated product developers wanting to put further investments on hold, or new team members with established relationships to a supplier, the process of the partner search involved technical, economic and social rationalities. The introduction of a new relationship (between Tine and Bremnes Seashore via the new researchers) and a new raw material (pre-rigor salmon) to the process changed the 'rules of the game'. This happened parallel to the above-mentioned shift in the organisation and framework conditions of the project. They could have considered other partner opportunities, such as aligning with actors closer to the market (e.g., distributors), but this was not seen as a significant pathway:

> Not when we considered alliances. Because then we had to go into the product, as we had big challenges in making the product work, stabilising it, and so we had a strong focus on this. (Hilde Torvanger, Tine R&D/marketing)

Stabilising production was considered a more pressing issue, and more strategic to the immediate progress of the project. Yet if, for example, access to distribution and customers had received the same emphasis in the argumentation as the access to raw materials, would then the process become very different? Anyway, Bremnes Seashore was in the end chosen as partner for industrialising Umi No Kami. In a note to corporate management at Tine, three potential partners were considered, ending with a clear recommendation:

> Choice of Bremnes Seashore is recommended for the following reasons: They have economic and financial strength, and a wish for collaboration. Lofotprodukter only wants production-for-hire, and has an overextended economy. Sunnfisk has an uncertain economic platform. Bremnes Seashore has 'super-fresh' pre-rigor raw materials, which is crucial to Umi No Kami. The others must get pre-rigor from other production units, and cannot do the processing industrially. (Note to the Tine management from the Umi No Kami project group, 2003)

Further, in this note a plan for 'gradual industrialisation in collaboration with Bremnes Seashore' was suggested, with a capacity for 200 tonnes in the production facility in the first phase. Behind this decision was, as mentioned previously, also a process of reorganising the biomarine portfolio of Tine, including Umi No Kami. I will now go on to describe the process of changing direction in the project in more detail.

The 'coup' of Umi No Kami

The process of reorganising the Umi No Kami project came out of the above-mentioned changes both in the corporate management at Tine, and in the management and merger of Tine Biomarin and Tine Ingredients. Swensen called this a 'coup' of the Umi No Kami project, to create more momentum towards commercialisation. A 'coup' is perhaps not an appropriate term, as the shift was largely initiated and implemented by Tine's management (the business unit director, Hovland, aligned with the new researchers and with the corporate management). Nevertheless, the shift represents a radical break with Umi No Kami as it had been originally organised and conceived of, and therefore the term helps describe the political drama inherent in the process. The result was a project changing from being clearly R&D-based to a commercial venture with ambitions to commercialise a brand concept. I had met up with the new constellation several times before I got to hear about this shift. In my field notes from the time period, we can read the following:

> During lunch at Tine R&D, the people around the table started talking about a person with strong opinions about the development of the fish salami, about some communication or collaboration problems, and about the fact that this person no longer had much direct involvement in the product development project. The contours of a conflict in interests came to the fore. It appeared that it was the identity of the fish salami (or of the Umi No Kami project), both in terms of technical specification and commercial presentation that were at stake. I learned that this person (Nordvi) had been central in the early phases of developing the fish salami, being project leader of the Neptun project and writing her PhD on the role of proteins in this process. I also learned that the project had changed so that it was not viewed as relevant to involve her anymore, but that she still had relatively strong opinions on the matter.

The fact that I had not heard anything about this during several weeks of my fieldwork triggered my curiosity, and I sought to retrace some of the events of this coup in subsequent meetings and interviews. The change started with the initiation of a reorganising process. Gunnar Hovland was hired as director for Tine Biomarin at a point when Tine had invested a great deal in blue-green activities but without getting anything back. Although a long-term perspective on these investments was being taken, it was time to start demonstrating some commercial potential. Around the same time, Swensen, Skjervold and the other researchers from the University of Life Sciences were hired by Tine R&D; Skjervold and Hovland soon found some common interests. Hovland interpreted his mission as being one of 'cleaning up the mess' – that is, structuring and organising the activities more efficiently, and evaluating what to do next. During this period, the initial Umi No Kami team gradually began to understand that something had happened, although they did not get the full picture:

> Bente Mogård was part of the team, and Hilde Torvanger, who had worked in marketing for many years, and they represented the market. But in particular Hilde was in a way duped out in that phase, and she did not know what happened at all, clearly she was forced into a tight spot. (Berit Nordvi, Tine R&D)

In not knowing exactly what would happen, they feared Hovland's scepticism regarding the whole project, that he would chose to close it down as part of his task to restructure Tine's biomarine activities:

> In the beginning, they were pretty sceptical to the product, which is perhaps not so strange, when they had a project that had not managed to produce anything during the entire fall of 2002. After a while, they prioritised the project, and they made the decision to only use salmon. From my point of view, there were not very clear indications that pre-rigor was so much better in terms of texture. But it was promoted as if it had very big advantages. (Berit Nordvi, Tine R&D)

Pre-rigor salmon triggered enthusiasm and new hopes for the Umi No Kami project in the new management of Tine Biomarin/Ingredients, and it was decided to remove white fish from the recipe, something the earlier project group had not been allowed to do. When the framework conditions had changed: 'They suddenly had a totally different set of conditions. Thus, they went for a new conceptualisation process, right, as they then only had salmon, they looked towards different markets (Hilde Torvanger, Tine R&D/marketing).

Why did the corporate management and board of Tine accept these rather radical changes in an already expensive and so-far unprofitable project? How could they accept changing the framework conditions for the project? They went from a blended white and red fish recipe to pure salmon, from doing domestic marketing prior to international marketing to then launching internationally first, and eventually from organising everything internally at Tine (R&D) to spinning the project out in a joint venture with Bremnes Seashore. In addition to changes in the top management and the business unit management, a part of the explanation came from the emerging impatience from the owners and the top management towards the biomarine projects. There was increasing pressure for commercial results, to produce some success stories to support the biomarine strategy, and for damming up for all above-the-line expenditures – for example, in Hovland's approach to his new job at Tine Biomarin/Ingredients:

> [We have] a lot of ideas about what we could do further. But Tine has until now not profited at all on such projects, even after having invested large sums of money. Therefore, these two projects[15] have to start generating profits before we can do anything more. (Gunnar Hovland, Tine Ingredients)

To Hovland, the way to generate profits was to create order within chaos and restructure the project portfolio:

> Now we need to show results on these ventures. We are in the process of checking for synergies between these ventures and Tine's core business. If we cannot find synergies, we will sell the companies, since then financial actors may do it better than us. Both the CEO and the head of the board are supporting and working hard to back up these activities. (Gunnar Hovland, Tine Ingredients)

With backing from the top management, Hovland now needed to prove the commercial value of the biomarine strategy before being able to expand. In a later meeting, Jan Ove Tryggestad could confirm my impression of the 'coup', and of impatience with the management regarding the Umi No Kami project. Tryggestad had been a long-time owners' representative on the board of directors at Tine. He had also been on the board of Tine Biomarin from the start, which, he argued, had not worked according to how it was intended. It became too research-based, and it was difficult to bring the projects towards the market. The impatience and uncertainty about the continuation of Umi No Kami had certainly become an issue in the top management at Tine. They felt that things had taken a lot of time, without showing enough progress and results. They were uncertain about whether and how the project should be pursued further. Hence, the shift was seen as a 'necessary change'. The board was impatient and frustrated over lack of progress. Tryggestad, as the owners' representative, fronted the project towards the owners and, as he said, there had been a 'long downturn' in representing these projects to the farmer-owners. Thus, this setting of a changed corporate management group, and Hovland and Skjervold's linking of Umi No Kami to pre-rigor salmon, as well as their renewed vision of commercialisation, convinced the corporate board and management that they should change their initial framework conditions to enhance their opportunities for getting something back on their investments on fish salami. The initial project manager for Umi No Kami, Rekdal, left the project for a job as managing director in another company, and the rest of his crew were more or less pushed to the side in this process: 'Gunnar [Hovland] had been hired then, and Per Olav [Skjervold] was hired too, and they started running their own course. Hence, it became a project with two groups that did not collaborate' (Berit Nordvi, Tine R&D).

According to several informants, the Umi No Kami project had lacked direction for a while before this, and the form, taste and quality of the product were contested issues among the different professional groups in the team. Nevertheless, during this period the technology and the recipe had been made to work, and developed into something more presentable. The groups involved were identified by Swensen as those working on (1) fermentation, (2) whey proteins, (3) process optimisation and (4) raw material quality/variation. There was a struggle for influence, partly by doing 'successful' experiments, partly via mobilising arguments and vision, and partly by shaping evaluation criteria. On the technical (experiment) side, the process of taking control of the project evolved through technical and argumentative trials:

> We started seeing good results on the effects of raw material variation, and we – who worked with this – did a coup of the project, and then the other group took the project back to their professional domain, and it has been different interests, where we have made coups of each others' experiments. (Lars Petter Swensen, Tine R&D)

At any given point in time, the group in control of the experiments had power to influence the further direction of the project, as long as their experiments showed progress in producing 'evidence' – that is, an improved product according to evaluation criteria that were also parts of the negotiations. In this case study, the battle was between representatives of pre-rigor salmon ('raw material variation') and representatives of bacteria cultures and whey proteins. Further, to strengthen their chances, supporting technologies were mobilised by the pre-rigor group. The efforts to measure (and therefore control) fat content in salmon intensified with Swensen's participation, even though it was the technology for this purpose was not feasible at that time:

> The reason I used saithe was that I knew the fat percentage in Norwegian salmon was very broad. Tine chose to sort the salmon with a technology they did not have, which won't make it easier to finish the project. (Erik Slinde, Institute for Marine Research)

Slinde, in delivering central premises for the 'bacteria and protein' group, preferred to control the fat percentage by mixing white and red fish in the mincer. Still, with Swensen and Skjervold, near-infrared technology was introduced to the project, first manually and later automated. When the experiments with pre-rigor salmon in the Umi No Kami recipe showed good results, it was not difficult to take control of the project and put one of the new researchers (Swensen) in charge on the R&D side.

> When we saw the significance of raw material variation, and then we also got a very good argument for leading and steering the other processes. It further laid the foundation for identifying the effects of the various types of fermentation cultures, and the effect of the whey proteins. (Lars Petter Swensen, Tine R&D)

By winning the battle for vision and leading the process, one also had the privilege, based on their 'model', of evaluating the effects of the other input factors. When the emerging new constellation gained strength, the story was effectively shaped on behalf of pre-rigor salmon as crucial for the quality and concept of the fish salami. The former lack of knowledge about the effects of input factors was seen as part of the problem:

> In that phase which I call 'chaotic', where no one knew the direction, did not have a product that held together, nothing, one just had a few research results showing a tendency, and so it was very chaotic because no one knew fully the effect of their input factors. Later, when we started getting a product that worked, it was a lot easier to gather the professional groups. (Lars Petter Swensen, Tine R&D)

It seems that the various input factors, such as knowledge and technologies from different professional groups, were assigned value post-hoc.

The effects were, in other words, negotiated among the participants, both according to the actual results of experimentation, and according to the story the leading group wanted to tell – in this case study a story of a novel raw material with great opportunity. In this quest for position in the project, the new researchers had worked 'on the side' of the formal experiment design to test their pre-rigor raw material in the Umi No Kami recipe and gain argumentative strength:

> The professional group dealing with whey proteins was very strong then, while Per Olav and I, we managed in a way to run over it. We did things on the side, like those things to do with fresh fish, supplementary experiments we did, that were not planned, but that we did after the others had gone home. (Lars Petter Swensen, Tine R&D)

In this way, Swensen and Skjervold could use informal activities to gain strength to formalise their position to realise their interests, interests that became aligned with the new management of Tine Biomarin/Tine Ingredients. But what other groups had been involved in the development of Umi No Kami? After buying the patent application from Slinde, Tine had cut off some of its relations to science and after this 'coup' even more science relations were removed from the project, while others were added. Swensen's account of academic input emphasised hygiene and chemometrics:

> How will the product stand a stress of, if you add a little listeria? And then this with NIR [near-infrared measurement], sorting on fat content? One group working on hygiene and one group working on chemometrics, or NIR. (Lars Petter Swensen, Tine R&D)

Research groups on these issues from the Food Research Institute and the University of Life Sciences had been consulted at various points. However, the knowledge fields of fermentation and proteins were left out of the story. According to the early Umi No Kami project team, valuable input had been made from such research groups at the same academic institutions. The post-hoc negotiation on the meaning and value of input factors also pertained to external relations.

In the revision of the biomarine portfolio, and the inherent evaluation of what to do with the Umi No Kami project, those who had been driving the project were not included. In a conflict on the direction, both on technology development and on conceptualising the product, the new management of Tine Biomarin/Ingredients did not open up their processes and emerging plans to everyone involved. Hence, some felt sidetracked and chose to withdraw from further engagements with the project, and turbulence in the team increased. After Rekdal quit, another three incumbants undertook the R&D project manager position before Swensen eventually got the job.

On the market side, Øyvind Kiland was hired by Hovland as commercialisation manager. At the same time the project was transferred to 'the line' to speed up commercialisation. The original project group consisted of dairy researchers (micro-biology), aquacultural researchers (technology and product development) and a chef, in addition to representatives from marketing and internationalisation. After participating in a meeting with the group, Kiland found it 'very easy' to downsize the staffing of the project. Swensen became project manager on the R&D side, as the only remaining permanent resource on the project.

> All others were just cut off. And we agreed to cut away most of the hours assigned to the project, and we reduced the R&D costs by two thirds. And it was easy to get approval for this, R&D and we were in the same boat, we had to succeed on fish. (Øyvind Kiland, Tine Ingredients)

For Kiland, lack of market interaction, at the expense of technical issues, represented a bigger problem in the project:

> I think that [Berit Nordvi] had influence on the group, and I know that there were internal disagreements. And I think she couldn't get [her suggestions] approved in marketing, thereby losing progress. I participated in one of these R&D/marketing meetings. It took long time, everybody wanted to say something, and there were around 10 people there, and there was perhaps just two that needed to be there at any given point in time. You could see people were tired, you could see signals between people. (Øyvind Kiland, Tine Ingredients)

'Endless' exploration of possible solutions to technological challenges resulting in loss of touch with marketing issues – that was how the early Umi No Kami process was represented by those taking over; they therefore took quite radical steps to reorganise the whole thing. However, in a reflexive moment, Kiland acknowledged the importance of *timing* in relation to distribution of roles and agency in this and similar projects:

> I think that if I had come into the project earlier, when the project had its thinking cap on, the patent application was there, and they were

supposed to make a product, then we would probably go through the very same idle development process, and I would have become synonymous with stagnation. (Øyvind Kiland, Tine Ingredients)

There were real (material) and challenging problems to be solved, problems of raw material characteristics, of technology development, combined with blindfolded explorations of market potential and appropriate conceptualisations of a product not seen in a market before. Such a complex process was perceived by Kiland as being so demanding and uncertain that it was necessary 'burn some people' in the process. Clearly, in this new phase, they needed different knowledge than in the earlier phases, but some of the earlier participants questioned whether the new organisation would have benefited from keeping closer relations to some of the previous work in the projects. Together with the sidetracking of people, a lot of the early knowledge and analysis of potential markets was also forgotten. Skjervold's expertise of salmon processing and his relations to Bremnes Seashore triggered Hovland and his organisation, and they started exploring opportunities for connecting Bremnes with the Umi No Kami project:

Gunnar Hovland was hired as the one responsible for the whole fish strategy. Per Olav had worked with the pre-rigor challenge at Bremnes, and of course he had a personal wish to cooperate, and that Tine should go for that company, then. (Berit Nordvi, Tine R&D)

A constellation formed of commercially focused management and pre-rigor salmon-oriented researchers. Hence, the people involved were encouraged to explore possible ways of connecting the two. In this way the technological interests of Swensen and Skjervold came together with Hovland's efforts to bring the Umi No Kami project towards the market, thereby showing some economic results to Tine's board and corporate management.

What was the 'need' in 'the market' for a fish salami? How should this project be shaped or changed in order to be accepted by distributors and convince potential users? While not having any clear answers available, the new constellation around Umi No Kami sensed that selling a product based on pre-rigor salmon would be easier and more valuable than staying closer to the initial idea of a mixed white and red fish recipe of fish salami. Hence, the argumentation for teaming up with Bremnes Seashore was strengthened. When arguing for the new project organisation, with Bremnes Seashore as partner, Hovland put a lot more emphasis on the *opportunities* that would open up by starting collaboration with Bremnes Seashore on pre-rigor salmon than on the challenges and costs of establishing a collaboration with this company. Crucially, they had also a letter of intent from a potential customer, a big French catering actor, which was used for everything it was worth in selling the new plan and mobilising resources:

> What in fact was the lever for industrialising the product was that Tine Biomarin had got a quite big and solid customer in France that wanted to use the product. Later, they have not been involved at all, but this was the lever that gave us accept for those investments. (Lars Petter Swensen, Tine R&D)

After having received the necessary mandate from corporate management, Hovland, Kiland and Swensen never mentioned the French actor to me again. The French had lost interest, but had been useful for the time being. I will now go on to explore some of reasons for connecting blue and green food industries, and what role this played in the emerging interaction between Bremnes Seashore and Tine.

The market system for fish

The idea of blue-green innovation seems to be based on a hypothesis that the connection of the two might help in overcoming some barriers and stimulate new industrial growth. As shown above, in the Norwegian public discourse on the topic, the fish industry was seen as representing a potential for growth through innovation and industrialisation, with its international market system and the recent and ongoing domestications of a number of biomarine species. However, the market system for fish was based almost exclusively on the exchange of raw materials, and the lack of industrialisation was seen as a main problem for innovation. Agriculture, on the other hand, was perceived as having little potential for growth, with its domestic orientation and the long-term decline in consumption of milk in the population. Still, the long-term development of academic and industrial knowledge within agriculture, and favourable economic and market conditions, had put the agricultural industry in a position of abundance. These differences, it was suggested, were complementary and a basis for synergies.

Tine's depiction both of itself and of the fish industry reproduced these narratives, supported also by the views of their academic partners, like Slinde and Skjervold. Hovland highlighted the lack of industrialisation:

> The problem in the fish industry is that they are poor at product development, innovation, marketing and branding. They might make some attempts, which almost always end up in the commodity shelves, the prices go down to the common level, and provides little returns. (Gunnar Hovland, Tine Ingredients)

Thus, there was little product development and little focus on marketing practices like 'branding' in the fish industry. The only actor in the fish industry that Tine found to be sufficiently 'serious' about innovation was Bremnes Seashore:

> Bremnes Seashore turned out to be the only actor in the fish farming industry that has pursued innovation; they have invested 40 million kroner on R&D since 1997, without really succeeding with it yet. All the others have no clue when it comes to innovation and research. (Gunnar Hovland, Tine Ingredients)

Although they had not succeeded yet, Bremnes Seashore was still seen as showing the right attitude towards innovation and industrial development. This picture might fit the reality at the time among fish farmers and fisheries pretty well. As mentioned previously, there had been lots of research and innovation, but mostly with focus on the farming process, with academic institutions and feed producers in leading roles. Less had been done on the production and commercialisation side. Bremnes Seashore, on the other hand, had for years been investing in new processing technology. Their competitors still had to wait through the 'rigor-phase' ('death-stiffness') after slaughtering the fish before being able to process it further:

> When all the energy in the muscle has been used up, decomposition begins, which makes the collagen break down, so that you can take away bones. This is what most processing actors are waiting for, and that is why it is as simple to transport the fish to France during the three or four days needed. (Jan Ove Morlandstø, MD, Salmon Brands)

This waiting time could be used to transport the fish both closer to international markets and to countries with lower labour costs. The hypothesis of the Bremnes Seashore people was that pre-rigor processing, eliminating the delay in processing and hence the advantage of sending the product abroad for processing, and in addition improving the quality of the product, would provide them with competitive advantage and better prices on their salmon. Succeeding with mobilising funding from the Research Council, and collaborating with university scientists for more than a decade, had led to groundbreaking methods for slaughtering and processing salmon. Yet thus far it had not led to any success in creating the expected added economic value. In this setting, one could easily think that there was little hope for industrialising fish, forgetting that, in fact, aquaculture, or cultivating fish, had been a great success story:

> In fact it has been going incredibly well, it is almost unbelievable. But that is if you look at the building of an industry. The last animal we domesticated was the turkey in the 1600s. To domesticate a new species, like salmon, is amazing. (Erik Slinde, Institute for Marine Research)

The first steps towards industrial production – or controlling the supply of raw materials – had been put in place. Still, the industrialisation of farmed fish had a long way to go yet, as compared with industrial production and marketing of agriculture. Very few fish farming actors had thus far made systematic attempts to develop industrial processing activities, and even fewer had gained any rewards for such investments. Bremnes Seashore, this relatively small fish farm and fillet producer, was one of the few that had tried. Together with the research group from the University of Life Sciences, Bremnes had, between 1993 and 2004, invested around 40 million NOK in developing technologies for slaughtering and processing of salmon, without any economic rewards within their established market system. In Bremnes' own brochure 'From Roe to Delicacy…Innovation and Quality All the Way' (2004) we can read:

> In 1993, Bremnes Seashore AS started a research-collaboration with The Agricultural University of Norway.…The project was successful and, with the new technology, Bremnes Seashore produces products with regular and higher levels of quality than other producers. This method, described as 'Cold Fish - natural cooling of slaughtered fish' is today patented, and recognised throughout the fish farming industry.

In other words, this was unique in the industry. These projects had led to patented technologies enabling industrial processing of pre-rigor salmon.

They had also led to scientific documentation of the beneficial effects of this processing technology on the raw material (see box 1). However, the patent, the scientific documentation of improved colour and texture, and the industrial advantages on logistics (fresher fish, improved durability) had still not paid off on better prices for their products:

> We have worked with [the ULS researchers] for some years to improve the quality of salmon. But we have struggled to realise those values when it comes to product development. We have had contact with many actors on the traditional side, but we can't get more out of it than the general price. (Olav Svendsen Jr, Bremnes Seashore)

Scientific documentation of superior quality obviously did not in itself qualify for higher prices in this industry. There seemed to be other issues at stake, in particular related to the way the market system was set up:

> We have a traditional marketing apparatus to traditional actors, like Halvard Lerøy and Coast, and they are really not very interested in these kinds of sales, because they only think of volumes, right, get out large volumes of filets to producers of smoked salmon. (Olav Svendsen Jr, Bremnes Seashore)

Their distributors showed little interest in marketing pre-rigor salmon for added value, as there were no market segments developed, and the

---

Box 1   Selection of scientific articles on pre-rigor salmon

Skjervold, Fjæra & Østby (1999) 'Rigor in Atlantic Salmon As Affected by Crowding Stress Prior to Chilling before Slaughter', *Aquaculture,* 175

Skjervold, Rørå, Fjæra, Vegusdal, Vorre & Einen (2001) 'Effects of Pre-, In-, or Post-rigor Filleting of Live Chilled Atlantic Salmon', *Aquaculture,* 194

Skjervold, Fjæra, Østby & Einen (2001) 'Live-chilling and Crowding Stress before Slaughter of Atlantic Salmon', *Aquaculture,* 192

Skjervold, Fjæra, Østby, Isaksson, Einen & Taylor (2001) 'Properties of Salmon Flesh from Different Locations on Pre- and Post-rigor Fillets', *Aquaculture,* 201

Skjervold, Fjæra & Snipen (2002) 'Predicting Live-chilling Dynamics of Atlantic Salmon', *Aquaculture,* 209

Rørå, Furuhaug, Fjæra & Skjervold (2004) 'Salt Diffusion in Pre-rigor Filleted Atlantic Salmon', *Aquaculture,* 232 fear of undermining their market for standard (non-pre-rigor) salmon was said to play a role. In such a market system, it was difficult to economise on quality improvements. High-quality and product development strategies were hard to implement. Both the distribution system and the marketing system were experienced as encouraging mass production of generic goods, therefore they had reached a point where return on their investments was required before making more investments of that kind:

> We may research and develop a lot, but you do that just for a while, until you actually get something back, get something commercialised, and that is the next step now, to prove something, and then further projects might come afterwards. (Jan Ove Morlandstø, Bremnes Seashore/Salmon Brands)

The Norwegian fish industry has 'always' been international, thus having far more experience with international business than agriculture. The trading of fresh and frozen fish with actors from the EU, Asia, Russia, etc. is the daily practice of Norwegian fish distributors, dealing with culture and language differences, and negotiating deals with various actors according to fluctuating market prices and demand, while often involving the Norwegian Ministries of Trade and of Foreign Affairs to get and maintain market access. However, the industrialisation of fish has only come so far in Norway, and most processing is done in other countries, such as Denmark and China. Two of the most common reasons for this, according to my informants, were the high labour costs in Norway and the added taxes for processed goods in the EU market. However, what about other potential reasons, like lack of R&D focus, lack of competence on industrial food production and marketing, and lack of technologies and marketing systems for more differentiated products? The marketing of fish was mostly done through 'idealised' market mechanisms, of more or less auction-based exchange, but with a rather poor ability to coordinate supply with demand, often explained by two conditions. First, the trade had been dominated by many small actors, thinking mainly short-term gains. This had started changing, as recently there has been a considerable concentration of actors in both in fisheries and fish farming. Second, the long production (breeding) time of salmon, around two years, created problems in unstable markets. Despite having some flexibility on when to slaughter the fish and further flexibility via freezing, there had been examples on how the whole industry had been led to crisis due to the combination of these two factors; small and vulnerable uncoordinated actors trying to breed and market as much fish as possible. Moreover, marketing was organised as an institutional-ised and uniform process of 'Norwegian' fish, by the Norwegian Seafood Export Council, with obligatory participation by the industry actors. This is 'generic marketing' through an image of 'Norwegian' fish. The implications of this model are threefold: less incentive for developing product variety/differentiation; potentially undermining trust in fish from Norway – because there are, in reality, quality differences; and less incentive for farmers/companies to improve on quality – due to compe-tition largely based on price. Yet, as previously mentioned, this sectoral problem seemed to be changing. Technical demands and developments were especially seen as speeding up this process:

> The consumer doesn't know where it comes from, he is seeing a fish, and only about 10% of dissatisfactions get back to the store. But when the big actors start putting their names on the product, the consumer requirements will rise, and the supermarkets focus a lot more on quality these days. (Jan Ove Morlandstø, Bremnes Seashore/ Salmon Brands)

If tracing systems were to become available to the users, and large pro-ducers started branding their products, the current practices of exchang-ing fish would be likely to change.

In sum, in this long-standing relationship between scientists and fish farm, they investigated various factors potentially influencing the qual-ity of farmed salmon, such as shape of fillets, impact of water quality, cooling of fish before slaughter, defreezing velocity, micro-biological growth and more. The emerging programme of industrial product devel-opment of salmon connected Bremnes Seashore, university researchers and, eventually, Tine. A constellation of marine researchers based at the agricultural university and a family-owned fish farm had done some-thing unusual in the fish industry: invested more than ten years and 40 million NOK in R&D. However, acknowledgement from scientific communities and public bodies is not enough in an industrial setting. Unless they led to added economic value for the industrial company, the projects were not considered successful. Hence, the search for part-ners that could help them commercialise and economise on the new knowledge led them through meetings with a number of actors in the established economic system for marketing fish, none of them show-ing interest or capacity for exploiting this potential. Then Tine arrived on the scene, representing an opportunity to escape the limitations of the fish industry, and to obtain a share of agricultural competence and infrastructure for product development.

The cold fish technology and pre-rigor processing

Increasingly, Bremnes Seashore came to be seen as the preferred supplier, and potentially also the production partner for the fish salami. But, what was this exceptional technology doing? By cooling down the fish in waiting-cages, the fish is less stressed during slaughter, hence slowing down the process of going into rigor mortis, and the fish may therefore be processed right away. If the fish is not cooled down, this process happens during just a couple of hours, making it impossible to split the fish and take away skin and bones in time. Thus, in normal processing plants, the fish would be stored for three to five days before processing. Following the excerpts from a sales presentation by Kiland in 2004, this 'cold fish' technology takes the salmon through the following steps (see Figures 12–18):

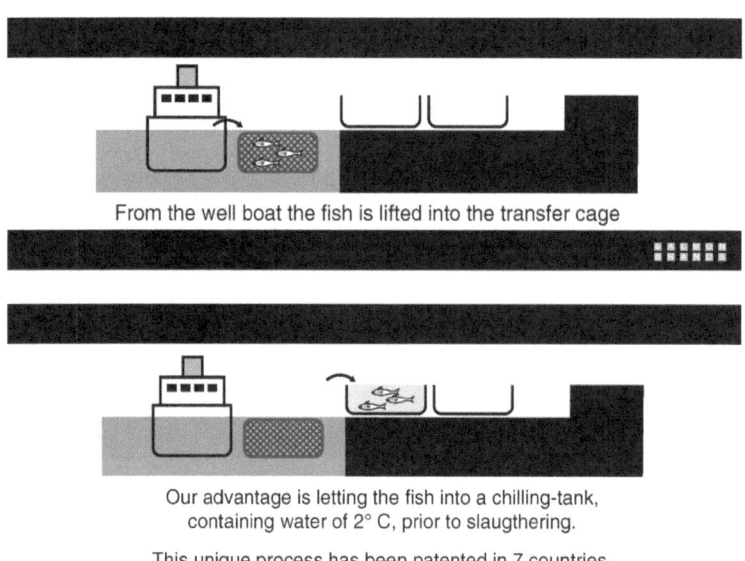

From the well boat the fish is lifted into the transfer cage

Our advantage is letting the fish into a chilling-tank, containing water of 2° C, prior to slaugthering.

This unique process has been patented in 7 countries.

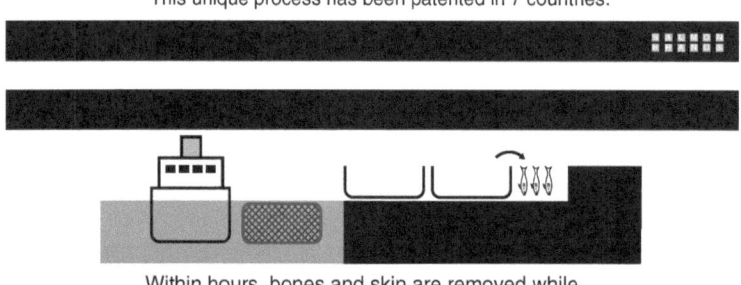

Within hours, bones and skin are removed while the fish is still in a condition of pre-rigor.

Giving several more hours for processing before reaching a rigor score in which it is impossible to technologically process the fish:

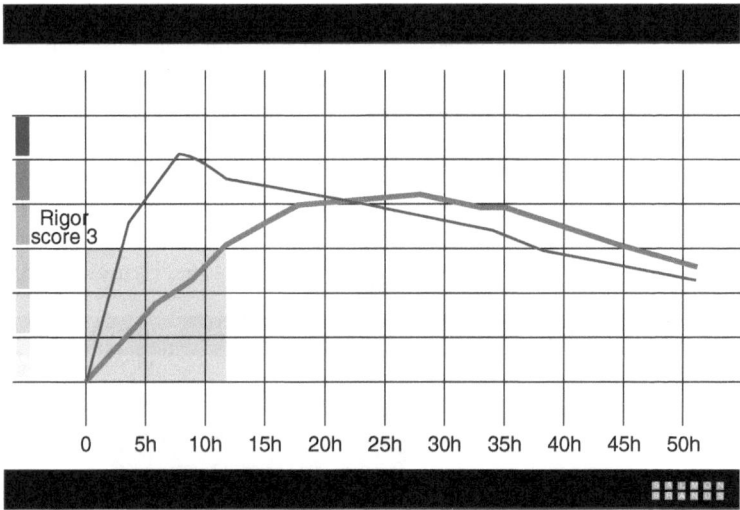

Through this technology, the UMB researchers have documented considerable effects on quality parameters of the fish. The colour is brighter:

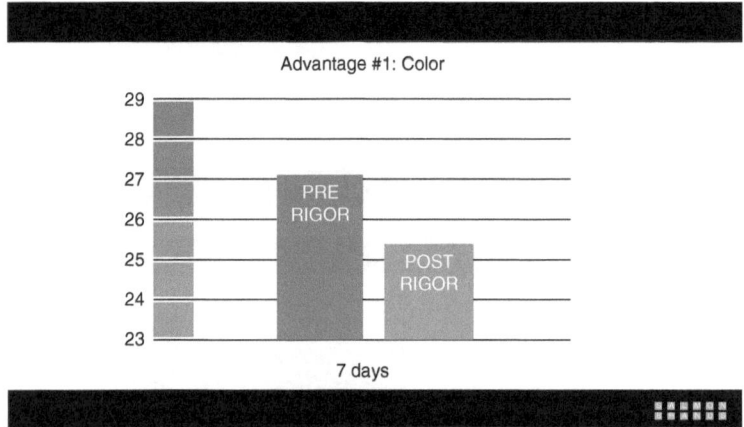

The texture is firmer:

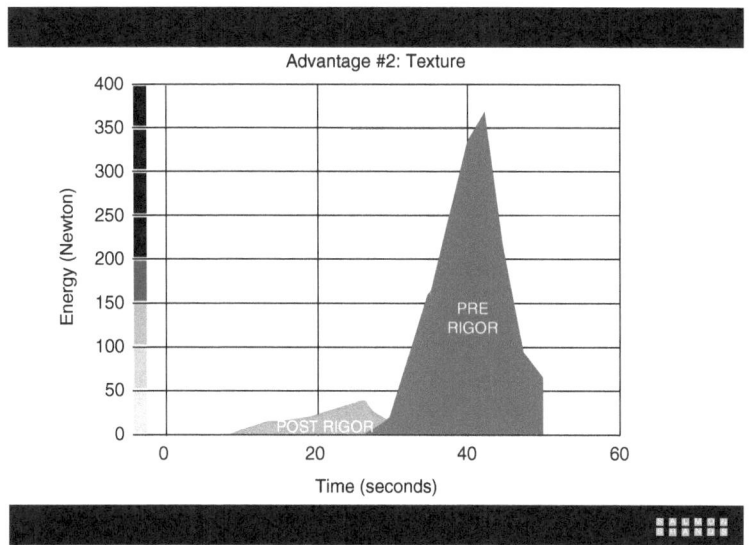

And there are fewer gaps in the fish meat:

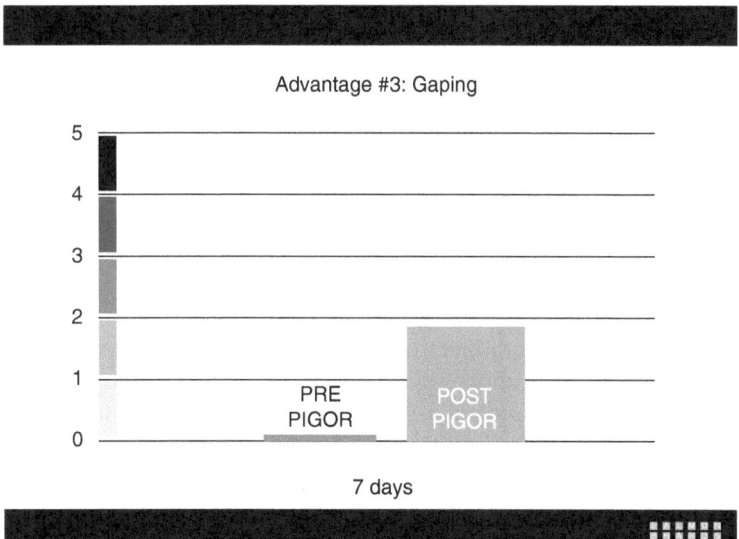

Figures 12–18   Presentation materials, Salmon Brands

As thoroughly documented in scientific journals, there was no doubt in Bremnes' presentation of its superior raw material. For Tine Ingredients, this became interesting for two reasons. First, the informal testing of this raw material as ingredient in Umi No Kami by Swensen and Skjervold had shown what they saw as very good and interesting results. Second, it was *experiencing* these qualities – seeing and tasting – that convinced management at Tine Ingredients that they not only wanted Bremnes as an ordinary supplier, but rather as some kind of partner in the project; it was a raw material and a technology with potential.

Bremnes Seashore's point of view

Why would a traditional family-owned fish farm, profitable and deeply embedded in the ordinary market system for fish, be interested in part-nering with an industrial actor from the agricultural sector? The simple answer is that they wanted returns on their innovation investments. A more nuanced answer is found in the evolving relationship between the two parties:

> Broadly speaking, we started with fermented salmon, and we were perhaps a bit flattered, because Tine had tried different kinds of raw materials, but it did not work out. That they came here and tested the raw material that we had worked hard to develop, and had some suc-cess with, were in a way important, we had a good starting point. So, it was a certain degree of enthusiasm when we went into the project, we found it very exciting. (Jan Ove Morlandstø, Bremnes Seashore/Salmon Brands)

The answer seems related to hope (for better prices on their fish) and flattery (acknowledgement of their technological developments); more-over, to their long-standing relationship to the researchers that had now been employed by Tine. Morlandstø emphasised the price issue:

> When you make a product like [the salami], it isn't that easily linked to the raw material prices by the customer, or the purchaser. They will not be that concerned with the current salmon prices at any point in time. And in addition, this product will not have any com-petition. (Jan Ove Morlandstø, Bremnes Seashore/Salmon Brands)

The hope to be able to produce added economic value from their high-quality raw materials was a clear rationale for starting the collabo-ration with Tine; an opportunity for Bremnes Seashore to get out of the 'fish industry trap' of raw material markets with fluctuating prices and little reward for product development. It was also interpreted as being a mutual need for help between the two parties, in creating value out of their respective innovations:

> We established a connection to Tine via Per Olav Skjervold, who had worked with us on various projects the last 10 years, and then Per Olav saw that fresh fish, pre-rigor, which we had been collaborating on, could be the solution to some of their problems. And, as we see it, Tine holds a huge competence, both on product development, and also on marketing and sales. (Jan Ove Morlandstø, Bremnes Seashore/Salmon Brands)

Thus, at Bremnes they saw this new relation to the big food corporation as an opportunity to get their product – pre-rigor salmon – out to customers in new ways that would provide better prices. Yet even with the advantage of building on established personal relationships, there were some challenges in establishing collaboration between the agro-food cooperative and the fish farm. When planning the production facilities, differences between the owner companies soon appeared:

> When we made a budget on machinery and equipment and facilities, they cut it down with a razorblade, and did it as cheap as possible. This is not exactly in Tine's spirit, right, we do things as expensive as possible, while they have cut costs, bought machines of too low quality, and some of them were second-hand and a bit worn. (Lars Petter Swensen, Tine R&D)

Different standards and attitudes towards equipment and costs created some frustration, as when second-hand machines had to be fixed and adjusted instead of investing in new ones. Different processing practices (standards, knowledge) also became visible, both in supplying raw materials of the right quality, and in processing routines in the production plant: 'Another thing is the fresh fish, where we also had problems with integration, to get them to follow the standards and hygienic guidelines we gave them' (Lars Petter Swensen, Tine R&D).

The change from a mixed white and red fish recipe to a pure salmon recipe also had implications for the cost of production, as this estimate over the mixed recipe shows:

> The raw material prices will vary according to type of fish, but e.g. 50% salmon and 50% saithe give nice products. The fish prices are at

present around 25 NOK per kilo for salmon, and around 5–6 NOK for saithe and fish cuttings. The water loss during drying will probably be around 35–40%. These costs in addition to the process costs will possibly lead to a total production cost on ready seafood sausage of around 40–60 kroner. (Project report)

From this example, we can see that production costs would approximately double by going for 100 per cent salmon, and this was before considering pre-rigor salmon. Moreover, this was not all; Morlandstø (MD of Salmon Brands/production at Bremnes) opposed machines for automating the near-infrared scanning. From his cost/use estimations, the machine could not be justified, unless the supplier took their part of the investment:

> I try to say that we are not in a position to buy an instrument for almost a million kroner, if we can't get more out of it than we do today. We try to control the fat content via the type of raw material we use, the size of raw materials. But clearly, if we had that instrument, we would have used it, and we could perhaps have used more types of raw materials. We can use it back on farming; provide information to the farmer on, for example pigmentation, thus they may use less [astaxanthin] in the feed, it is a very expensive part of the feed. (Jan Ove Morlandstø, Bremnes Seashore/Salmon Brands)

Thus, on the one hand, Morlandstø did not feel they really needed an automatic near-infrared instrument, as the cost could not be justified. On the other hand, it would have been very useful if they had one, and he suggested that financial partnering with the fish farm could possibly justify the investment, as its use would have served several different interests and purposes at the same time. In total, the cost of partnering with Bremnes Seashore and using pre-rigor salmon instead of a mixed recipe were considerable: investments in a new production plant, development of new technology to control fat levels, work on production routines and especially the doubling of raw material costs. These were investments and costs that the new constellation at Tine were prepared to take, but that – paradoxically – the new partner, Bremnes Seashore, tried to get as low as possible, at least on the equipment side.

Partnering: negotiating contracts and practice

The differences between aquaculture and agriculture related to 'culture', competence and market systems were frequently mentioned by various participants in this case study. However, when it came to collaboration between researchers from Tine R&D and the processing staff at Bremnes, communication seems to have been working without much frustration. This did not mean that differences had disappeared, but rather that some communicative features seem to have been in place between these groups:

> [Bremnes Seashore] have collaborated with ULS since the early 1990s, related to developing pre-rigor processing and slaughtering methods. Both improving effectiveness and getting that quality level they have today. So in that respect it is much easier for R&D people to be accepted over there. (Gunnar Hovland, Tine Ingredients)

The (partly) common history of collaboration between Bremnes Seashore and the University of Life Sciences had produced personal relationships, new knowledge and technology, and probably a common language about the new practice. In addition, the division of labour between the Tine people was claimed to be important for providing a good collaborative climate. Technologically, Swensen (Tine R&D) worked with the Bremnes processing unit as usual, in close dialogue with Kiland (commercialisation manager, Tine Ingredients), while Hovland took care of the partnership negotiations. A year into the joint venture, but still negotiating issues of organising the joint venture, Kiland described their roles:

> It has been very important for progress that Gunnar has been working on this, and then I am only working on my things. If I had started arguing with them over money, this would have been ruined, because I can't argue with them over a million or ten on Monday, and then ask the production to provide a test production for me in Tuesday. And Gunnar is consistently competent, and very good at selling. And, as I interpret it, he had the [Tine] corporate management in the palm of his hand. (Øyvind Kiland, Tine Ingredients)

Thus, having Hovland to take care of overarching negotiations with Bremnes, and also provide support and resources from top management at Tine gave freedom for the 'operational' project participants (particularly Swensen and Kiland) to focus on their particular tasks, furthermore making it easier to collaborate with Bremnes Seashore on operational tasks.

The project management and Tine Biomarin/Ingredients sought continued support for the project from the Tine corporate board and management. In a project report (December 2003), we can see how the various arguments were combined to strengthen a clear proposal for taking the project over to a commercialisation phase. Six man labour years should be assigned to the project. After investments of 16.5 million NOK in the UNK project, they were 'now ready for industrialising and commercialising this knowledge', hence asking for another 12.2 million NOK the next year (2004). With regard to conceptualisation and market, Umi No Kami was argued to be a global concept, partly because of its global novelty – 'a real innovation' – and partly because of their international scope regarding marketing. It was supposed to be a 'gourmet product for restaurants and delicatessen, ingredient for food industry (salads, sandwiches, etc.), weekend sandwich filling for supermarkets, hotels and cafés, and a snack for supermarkets, bars and restaurants'. The sales estimate was set to 1,200 tonnes a year in 2010, and perhaps 5,000 tonnes as the peak (long term), 'providing both national and international work in parallel' (Project report, Umi No Kami, December 2003).

Presentation materials from a meeting between Tine Ingredients and the management at Bremnes, from the initial dialogue on opportunities for formalising collaboration, displayed how this was communicated to their future partner on the project. The aim was to repeat Tine's international success with Jarlsberg cheese – the most widely imported cheese in the US, and with an increasing demand in a number of countries worldwide. UNK was said to be presenting a 'new and delicious way to eat salmon', with high quality, developed in 'collaboration with top class chefs'. Market research was in process, awaiting results regarding market potential in the US, Asia and EU, and the planned 'entry strategy' would focus on 'selected segments offering high-quality and high-value positioning'. Collaborators would be 'high profile', and the first feedback from Nutrimer, a large catering actor in France, was that the product gave 'a good mouth feel, in line with luxury delicacies, not making the sandwich wet or greasy'. It was viewed as natural with positive micro-bacteriological properties, and suitable for salads, pasta dishes and sandwiches. Their requirements were: no additives, stable quality and six-month self-life, positioning the product in the luxury segment with estimated volumes of 30 tonnes for sandwich and 200–400 tonnes for salads for testing during 2005. If an agreement could be reached, production would start in June 2004, with the 'Tine board and management fully behind the project'. Three phases of scaling this up were outlined: first, pilot production, then common development project, before industrial production. During the pilot, Bremnes needed to invest 4 million NOK in facilities, while TINE would put 12 million

NOK into marketing. Preferably, TINE wanted a joint venture, with Bremnes having 34 per cent of shares (Presentation to the Bremnes Seashore management, 30 January 2004).

In a Tine corporate board meeting in April 2004, a draft for a shareholder agreement between Tine and Bremnes Seashore was presented. Bremnes was presented as producing 14,000 tonnes of fish per year, and as the owner of two patents: 'Process and plant for handling of fish from its delivery to the quitting thereof', and 'Method for the manufacture of a high-quality fish product' – in other words the mentioned patents for live cooling and pre-rigor processing of salmon. Tine had the rights to the recipe and the patent application for the fermentation of fish, a product that had 'particular requirements to production process, based on Bremnes' slaughtering technology, covered by the above-mentioned patents and patent application'. The investment in the joint venture was said to be up to 30 million NOK during 2004–7. In the statutes, the purpose of the joint venture, called Salmon Brands, was formulated as 'increased value creation on fresh salmon through marine innovation, processing and branding'. The company's activities were to be 'R&D, production, brand development, marketing and sales of products based on fresh salmon' (Board meeting case papers).

In the Tine board meeting (27 April 2004),[16] the management presented Bremnes Seashore as motivated and competent, emphasising Bremnes' considerable investments in innovation the last decade, hence recommending the partnership. Some board members expressed the fact that they were fed up with dealing with costly blue-green projects, but perhaps more as a preliminary warning than as an attempt to stop the project at the moment. The optimism of the project management was taken a step down, but the opportunity to utilise more of the surplus of whey was still seen as an important factor. The corporate management representatives got support from some board members for the suggested way forwards for Umi No Kami. In the end, the management finished the discussion by signalling both modesty (adjusting sales curve and admitting tough aspects) and strong optimism (would have invested themselves). The case ended with full agreement to the suggestion of carrying on with establishing collaboration with Bremnes Seashore, and Refsholt, Tryggestad and Hovland were appointed to the interim board of the Salmon Brands project.

In order to build the argumentation for a final decision at the next corporate board meeting, a report was written by Tine Biomarin/Ingredients about Umi No Kami, and distributed as attachment to the board documents. The report is dated 18 May 2004, and consists of a summary of the

Umi No Kami project so far, its present status and an operationalisation of the next phases of production and marketing (see excerpts below):

Status

Patent application

- The best protection against imitation is to develop a strong brand based on a good concept and then continuously developing the product so that we are always one step ahead of those wanting to copy it. This philosophy is valid both with and without a patent.

Industrialisation

- Pilot facilities are being established at Bremnes. Project plan implies test production in June 2004
- Technology investments are minimised by the purchase of used equipment at the start, until we have documented repurchases of the product.

Sales/market development

- Expecting test sales and test on repurchase during 2004
- Norway: TINE Ingredients has started testing at a customer. TINE Storhusholdning [catering supply] is awaiting test production and can thereafter target selected customers for test sales.
- France/Nutrimer: They have received the product, and communicated back that the quality is within the target area, but that it is too expensive. Further, they claim that substitutes exist. We interpret this as a negotiation move. The presumed substitutes are ordered.

Production costs

The company's production calculations were presented at the board meeting, at 110 NOK/kilogram. The elements of risk of this calculation are tied to:

- Raw material prices (based on 30 NOK/kilogram, has varied from 14 to 37 during the last 5 years). In the production contract with Bremnes, the pricing will be market based, but locked in a certain interval.
- Loss (50% of the salmon is not used in the product – hence double raw material price). This can be improved by further product development of relevant minced products (hamburger and similar).

- Test production. Minimal sizes on test batches down to 100 kilogram/batch, equals 11,000 NOK per batch.
- Water content/activity. The product is based on drying salmon from water activities at 90% down to 40%. It turns out that the product provides a better taste experience with a somewhat higher content of water. Every percentage increase will positively influence the calculation.

(Report for the corporate board, from Tine Biomarin, dated 18 May 2004, attachment to board meeting documents)

Product development was said to be progressing towards the point where the product would be technically feasible. The resistance from Bremnes towards buying new and expensive equipment had been integrated as an advantage in minimising investments. In hindsight, it seems that the French actor, Nutrimer, was ready to discontinue the relationship, but was still used as a representative for the market potential of the product. Further, they showed concern for reducing the costs of production, both in the test and up-scaling phases, and in regular production. Based on this report, the final decision on the new business strategy and partnership of Umi No Kami was discussed at the next board meeting on 26 May 2004. When introducing the case to the board this time, the management stressed how they had taken suggestions from the previous meeting into account. Further, they told about the rejection of the patent application, the fact that production facilities were not yet in place and the challenge of high production costs. Then a three-staged future process was described, each with exit opportunities if the project did not demonstrate satisfactory progress and results, before ending by recommending the plan. In this way, the project was presented as vulnerable and with some risk involved (although some of the risk had been reduced).

The response from the board members was immediately supportive and, after some discussion, the recommendation was approved by the board with some small adjustments in delegating conditions. While some board members still expressed a degree of scepticism, the majority had now become clearly positive to the plan. It was also argued that Tine had now learned from their mistakes, and was thus more competent to do biomarine projects than before. Finally, the taste was mentioned as now being 'good', meaning that they had now gained control of the recipe and technology, enabling stable production of a good-tasting product. In sum, Tine, their partner and the object itself were ready for commercialisation. The discussion was closed by the management again modifying their ambitions and reassuring the board by referring to their ongoing market research and by making clear that this task would be taken seriously.

In having acquired the necessary support from Tine's corporate management, negotiations with Bremnes Seashore could be formalised and intensified. The people at Bremnes were sceptical to the initial proposals from Tine, and Hovland explained how they worked both with direct negotiations and informal dialogue:

> There was a lot of arm-wrestling about how things should be. They had to integrate many other interests in the agreement, on both raw material prices, on growth – capital for growth, and other things, where of course Tine held the opposite point of view. And we used the chairman of the board and the CEO of Tine, to, you could say; simply make them familiar with Tine's basic values. And gradually as they trust us, they are willing to go further with us as an actor, instead of others. (Gunnar Hovland, Tine Ingredients)

Having different interests and being unfamiliar with each other was gradually overcome by exploring and getting to know each other, and concurrently negotiating how the different aims could go together. Not only sectoral differences, but also organisational and historical matters had implications for the negotiations. Representing Bremnes, Morlandstø spoke about the need for them to achieve a larger change of the agreement:

> In the beginning the agreements looked more like a rented production deal, but we were absolutely not interested in that. We have turned these agreements upside down these, and now we have a sensible basis for both parties. (Jan Ove Morlandstø, Bremnes Seashore/ Salmon Brands)

It was important for Bremnes Seashore to be an equal partner, more in terms of status than owner share. After six months of exploring each other and negotiating the terms for a joint venture, the first formal steps towards commercialising Umi No Kami as a pre-rigor-based salmon salami were settled, and the joint venture Salmon Brands AS was established. Tine's risk and commitment for the project had been shared with another company, which was rich with technology and raw materials, but without industrial experience beyond the ordinary market system of selling salmon as generic non- or half-processed raw material.

Increasingly, through the process of scaling up production, the Salmon Brands organisation felt a need to have more control of the raw material side of the production process. This strengthened their argument for including all the pre-rigor activities from Bremnes. Jan Ove Morlandstø, having become managing director of Salmon Brands, was concerned about his influence on the raw material quality as rationale for this expansion:

> I don't think there was any other way to do it. If I am going to buy these products and this quality without having the power and authority to change it, it will fail. (Jan Ove Morlandstø, Bremnes Seashore/Salmon Brands)

These negotiations resulted in moving all processing, pre- and post-rigor, from Bremnes Seashore to Salmon Brands. Suddenly, the joint venture had considerable volumes going through their system, without having closed a single commercial deal on the innovations of Umi No Kami. The question was about how to draw boundaries in the joint venture, between further processing and marketing, or between raw material supply and processing/marketing. The actors ended up sharing more in the joint venture than initially planned, solving potential controversies by sharing both costs and income within the whole industrialisation process, hoping it would mobilise mutual motivation and commitment to the joint venture:

> To get better prices for the salmon than just the raw material price, you have to do something with it. Usually we make salmon filets, right, but its price is related to the raw material price anyway. And so, they had this idea, and they had a very good sales apparatus, and plans to sell very large volumes of the sausage then. (Olav Svendsen Jr, Bremnes Seashore)

In having settled an agreement for doing product development, production and marketing together in a joint venture, what remained were largely two issues: scaling up production at Bremnes, and marketing the product. Even if these two processes partly evolved parallel to each other, I will start with the up-scaling of production, before I move on to the marketing process in the subsequent section.

Scaling up production

When moving production from the laboratory at Tine R&D to large-scale facilities at Bremnes, problems arose in the relations that existed between micro-biology, production facilities and knowledge/routines. In addition, the high-quality demands of the product became even stricter in interaction with market representatives.

On 28 June 2004, it was time to start scaling up production at Bremnes, and I joined the Tine team on the trip to the west coast to observe the event. For two days, people from Bremnes and Tine worked together on preparing and starting first large-scale production in the new production facilities, with training personnel, adjustments of technology and intervention from micro-organisms at stake. I met up with the Tine team at the airport early Monday morning. It consisted of Lars Petter Swensen (R&D project manager), Svein Erik Hilsen (chef), Tore Teigen (consulting sausage maker) and Dorotha Dynda (micro-biologist). New production facilities were supposed to be ready, composed of partly new equipment and partly used equipment from various Tine dairies. At the airport, Teigen, an experienced sausage maker hired for the event, told Swensen about his scepticism regarding the status at Bremnes, stating, 'I am ready, but they are not ready over there.' Here, he was commenting on what he saw as lack of planning and overview, particularly some technical issues that had not been solved yet. Swensen responded in a relaxed mode, 'Well, the most important thing is just to get things through this period of time; training.' After the flight from Oslo to Haugesund, we drove a rented car for two hours before arriving in the small coastal village on the south-west coast of Norway, where we soon found out that not everything was ready. In fact, it was quite chaotic. Some technicians were still working on the drying and smoking facilities, while a supplier of a production machine was installing software and trying to get the thing ready for its task.

From a presentation of the status of the Umi No Kami project to the Tine management by Swensen, we can make a summary of the (technological) achievements and remaining challenges of the project by the end of 2003, just before the processes of conceptualisation and production were speeded up during the spring 2004. The goals for 2003 had been to make two different recipes ready for production, one 'gourmet variant' and one 'grits variant'; after ten experiments, the grits variant was ready, but the gourmet variant still had some challenges to do with binding the fatty acids. The project claimed to be pioneering work, as 'there is little knowledge on fish and minced meat of fish worldwide', developing core competences on raw materials, fermentation, product, process, design, protein and emulsion. The year 2003 had been used for narrowing the scope (to two specific recipes), by working on a set of technological issues. In order to control the raw material, and thereby achieve better stabilisation of fatty acids, use of NIR had been tested with the aim of standardising the percentage of fat in the salmon input to 10 per cent. Furthermore, the use of fresh instead of frozen raw material (salami from meat is produced from frozen materials) was concluded to give a better texture.[17] The final choice of using only salmon instead of a mix of red and white fish was presented as having technological advantages, as the 'access to raw material and control is much better'. Whereas tests using different parts of the fish had not demonstrated significant improvements, and the use of 24-hour-old raw material only had provided minor improvements on smell and taste, use of three- to six-hour-old (pre-rigor) raw material was considered to provide major improvements in smell and taste, without needing to add transglutaminase (an enzyme), and giving better shelf life to the product. The remaining R&D challenges for the year to come (2004) were said to be to document shelf life, improve fat binding in the gourmet variant (by 'changing process conditions and reducing the percentage of fat in the raw material'), scale up production at Bremnes, expand the product variation (shapes, spices, etc.) and, finally, develop packaging. While improving the fat binding in the gourmet variant did not present big problems, scaling up production brought some challenges.

Back at the start-up event at Bremnes, a meeting with the management and production personnel had been scheduled. Eight persons from Bremnes Seashore and five from Tine were present. An experienced dairy manager had joined the Tine team from their western region. Swensen started out with a brief version of the story. Then, the sausage maker, Teigen, took over, and in a narrative style he told stories of how meat was traditionally dried and fermented, as way of extending the durability of the food. Next, he compared this with the fish salami, on levels of salt and fat, the recipe, drying and smoking, and so forth. Then he warned the production workers against mould. Calcium sorbate might be used as an anti-mould ingredient; as the product might mould overnight, it would look like snow in the entire room, he warned, and if something white was detected, they would need to clean the room immediately. Mould might also influence the flavour of the sausage. In moving on to bacteria cultures, now being produced industrially and added to the recipe, Teigen related how, before the 1970s, this was not something that one 'made'. Starter cultures were just something residing 'in the house' of production – remaining from the previous production; for example, in the liquid solution, represented by an almost magic understanding of the phenomenon. A couple of 'failure stories' were then provided to underline the importance of the bacteria for a

proper fermentation process. Swensen then went through the processing work of the raw material:

> As fish varies in size from 2–7 kilograms in the same brood, we are
> trying to develop a system for continuously measuring fat content
> online. All brown fat[18] has to go, and we will use only back loin, since
> it is leaner. There is a close interaction between raw material, ingredi-
> ents and technology, and quality is extremely important.

Some of the production personnel responded a bit anxiously, though
also jokingly, that they felt this was a big responsibility, whereby
Swensen calmed them down by arguing that this is 'dairy production, which we have been doing for 100 years, we will do fine'. In the
second session of the meeting, Dynda, micro-biologist, took the participants through some basic issues of micro-biology, such as shapes
and clusters of bacteria, how bacteria reproduce, how bacteria tests
are done, etc. It was matter-of-fact information from her professional
knowledge.

The next morning (Tuesday) was used for a course in hygiene, run by
the dairy manager from Tine West. In this way, Swensen and the others
had time for getting the facilities ready for production. Programming
and adjustments of the machines still remained, and they had discovered that they lacked the right type of cleaning agents for the drying
and smoking facilities. Production personnel were taught about the pas-
teurisation of milk and how most bacteria thrive at 20–40 degrees, and
survive cooling, but die from temperatures between 72 and 100 degrees.
'But then', he said, 'we are not only working against micro-organisms,
we are working with them too, facilitating the best possible conditions
for their development.' Bacteria were participants in the project, it was
crucial to their success that they acted as they were supposed to and it
was challenging to succeed with the recruitment, development, control
and good treatment of them.

Back on the production premises, they could not get the vacuum
pump of the blender to work, and instead chose to dismount it, running the first round without. Somewhat resigned to the situation,
Swensen explained the mess by the cultural differences between Tine
and Bremnes Seashore: while Tine was thinking long-term, Bremnes
went for fast changes and adaptations. At 9.50am, Hilsen, the chef,
hosed down the facilities, while Swensen gathered together the last
pieces of equipment before the training session started. The production personnel were taken through the basic workings of each

machine, discussing important things to remember and possible ways of routinising this new production practice. Dynda and the dairy manager ran through another round on bacteria culture, lab testing and hygiene, before the cleaners arrived. Everything was improvised, in the hopes of being able to produce something before the Tine team had to leave for the airport the same afternoon. Finally, after cleaning, things were ready, and the first test production on the new large-scale facilities was successfully done. Though it was both chaotic and calm at the same time, a lot of people passed through the production hall, some just observing and checking on the progress being made there, and others being more or less involved in the event, all with appropriate sterile white clothing, caps and clogs. The resulting product was a pink salmon 'salami' in opaque black sausage skin, to be hung in the drying and smoking facility to mature during the next few weeks.

The intervention of micro-organisms

The transfer and up-scaling of production from the dairy lab to the large-scale facilities at the fish farm was undertaken during June 2004 and everything seemed to work well, but then, in the following batch in July 2004, problems occurred. Suddenly the whole batch of salami was attacked by mould and had to be discarded. Overnight the harmony was gone. This was also challenging for the marketing people who had already started presenting the product to various international customers. Just a few weeks before the first 'marketing tour' with Salma, where they planned to visit a number of contacts worldwide, the little devil of a micro-organism suddenly covered a whole batch of salmon salami. A white layer of mould had invaded every salami in the drying facility, threatening Salma's first public performance. It was a production worker, Magda Sæverud, who discovered the mould:

Sæverud: It was mould, and we washed down the sausage four or five times, but in the end we had to dispose of it, everything, a whole production.
Researcher: What was the reason, do you know?
Sæverud: No, I don't know. In reality we don't know, but I have come up with some thoughts about it. We used this machine for making, we called it b-sausage, and I think that it perhaps was not clean. Because the mould started on that sausage, and then it spread from there.

This explanation regarding inferior quality and testing a machine for its ability to take away bones, resulting in it not being totally clean was supported by Sæverud's line manager, Bjørn Rino Jacobsen:

> We produced among other things a sausage with raw material that had been through an old grinder we have for separating bones, and the moulding of the sausages started on these sausages: Something in that machine that could have infected the raw material. (Bjørn Rino Jacobsen, Bremnes Seashore/Salmon Brands)

Yet their managing director (of Salmon Brands), Jan Ove Morlandstø, had another opinion of the problem. In his view, it was not particularly the use of the grinder, but rather failed routines that could explain the event:

> It is clear that we should have nailed our routines better, and we had a plan for improvement when this happened. Some of our equipment is old, so if the power goes off, the programme resets, and it happened twice with that production, so the temperature control went wrong. We have added routines having people watching things at night and in week-ends, so, I hope that this won't happen again. (Jan Ove Morlandstø, Bremnes Seashore/Salmon Brands)

Swensen had to immediately take a flight to the west coast and try to get in control, find the reason and make sure that it didn't happen again. The technologist went through several new rounds in teaching and controlling the production workers at the fish farm to secure a more stable and predictable production. Eventually, they concluded the reason was a lack in the maintenance routines of the machine and possibly not being strict enough on the routines on hygiene. Even though both issues were described in the work manual, they had obviously not yet been established as stable practices by the local workers. In addition, Morlandstø – the strongest opponent to investing too much in new and expensive equipment – admitted that old equipment might have contributed to the problem. After this event, through the reinforcement of new standards and routines, the problem disappeared and did not reappear in later productions. However, this problem had troubled Tine R&D previously, for around six months, early in the Umi No Kami project. Interestingly, in spite of Teigen's reminder to the production workers about the potential problem of mould, it reappeared when moving production from the laboratory to full-scale production facilities. Yet after the problem had occurred, the project management showed the ability to mobilise quickly, and succeeded in removing the problem.

The activities of marketing, developing and producing were intertwined throughout the process and particularly during the scaling-up of production and the parallel work of conceptualising and presenting the product to potential users. Due to economic pressure, Kiland, the commercialisation manager, chose to test-launch the product before they had any guarantee that the transition from the laboratory to large-scale production worked out well. To produce (test-) batches without *using* the products for commercial purposes would rapidly increase the costs beyond what would have been viewed as acceptable, and to have production running without customers would in any case be perceived as demotivating, both among production workers and management. Many things could have gone wrong and, according to the project management, the only way to succeed in stabilising all these factors was through trial-and-error learning. When I challenged the commercialisation manager, Øyvind Kiland, on this issue he seemed cool, trusting his colleague's competence in dealing with the issue:

> Yes, well, I was very comfortable with this. Because I had great trust in Lars Petter, and trust is everything here, feeling that, he had Ommund[19] up in Bergen helping, it was people I trusted. (Øyvind Kiland, Tine Ingredients)

This trust in Swensen and his colleagues' ability to solve their tasks together with Bremnes did not fade during the process, even in the face of more technical challenges as the commercialisation process evolved.

Adjusting production practice

To what extent did this scaling up of production across settings demand adjustments and the change of existing technologies, or the learning and adaptation of new technologies? Some of the problem with scaling up the pre-rigor production had to do with the rest of the processing going on at Bremnes. To rigidly prioritise the raw material needed for the salmon salami in large volumes would demand Bremnes Seashore to turn the order of pre- and post-rigor production upside down. In the beginning, post-rigor was done first in the morning, due to significantly higher volumes, and then pre-rigor was done afterwards. Still, this was unacceptable for securing the quality and sorting required for salmon salami production, hence interfering with the organisation and logistics of post-rigor processing. Further, although pre-rigor was of strategic priority to both

Salmon Brands and Bremnes Seashore, in practice, the economic value of the well-established business of post-rigor production had its influence. As mentioned in the section about contract negotiations above, we saw that the solution to this dilemma was to include all Bremnes' processing activities in the Salmon Brands joint venture, hence increasing Bremnes' owner share to 49 per cent, and giving the Salmon Brands management full control of the processing priorities and practices.

While waiting for the development of automated NIR-technology, and discussing whether and how to finance it, they had to sort the fish partly by using a narrower range of fish, sorted by weight, in addition to scanning the fish manually:

> Our machines are best, most efficient, on three kilogramme fish and upwards. Salma should preferably have two to three kilograms, and three to four. But if we use three to four, we have to cut off more of the belly, because it contains more fat. (Bjørn Rino Jacobsen, Bremnes Seashore/Salmon Brands)

If successful, the production volumes would then be increased radically beyond the initial 'training levels'. This way of sorting would lead to larger problems both with manual scanning, manual processing, getting enough fish and getting a lot of by-products – cuttings – increasing the raw material costs, if not also finding use for these materials:

> Until now, we have cut off what we call 'belly mouldings', we have a product in this. But the largest challenge production-wise is to get the job done fast enough. It is a complicated handcraft, as it is pre-rigor fish, and you have to get all the bones out, and all the brown fat too, and today we have no machines doing this. Should we invest in machines and equipment, and do we have the necessary space? (Bjørn Rino Jacobsen, Bremnes Seashore/Salmon Brands)

According to Morlandstø, this was not too much of a problem, as they sold 'huge amounts' of belly mouldings to the Japanese market. The question of scale was both about production capacity and sales, and the need to balance these in a flexible way. The ambitious production targets could even have made problems for ingredients supply and production capacity, but, on the other hand, these targets were admitted to be unrealistic:

> Salma, just on the Norwegian market, was to have volumes somewhere between brown cheese and mackerel in tomato sauce. And if you think about brown cheese, which I know well, it is about 9000

tonnes, it has such a volume that the next big discussion in the group, and we have laughed a lot about this, was about whether there was enough whey in Norway, because it contained a few percents whey. (Øyvind Kiland, Tine Ingredients)

Nevertheless, achieving a decent production capacity not only meant dealing with technology, raw materials and ingredients; it also required lots of training of the production personnel, often in combination with technology adjustments. I asked Sæverud, a production worker, on the combined issue of learning and technological adjustments:

Researcher: How much are you capable of producing per day now?
Sæverud: Well, I think it is around 500 kilograms. If everything goes as it should, but we have had problems with that sausage filler every time now. Because the sausage skin breaks, and so the next time I think the supplier will be with us. It creates some waste.

Handling fragile raw materials and learning to use new machines took both time and effort, and here we saw how the machine supplier had to be brought in to participate in solving the problem of breaking sausage skins. Jacobsen, line manager in the Salma production, explained more about the handcraft skills needed to handle the process, and the possibility of adding technology:

It is very time consuming work, making the raw material needed for the sausage, right. And it is supposed to have this and this content of fat, and then we have to do the things I referred to of 'hand-craft', taking away brown fat and all that. As I see it, we need new equipment to do the job, so that we can do it more effectively than today, everything is hand-crafted. (Bjørn Rino Jacobsen, Bremnes Seashore/Salmon Brands)

In this work-intensive process, Jacobsen suggested rethinking the distribution of labour between machines and humans to increase the productivity and simplify the job for his team. Another question was, of course, the cost of training; of producing a product that had not yet found its commercial users (i.e., a 'market'). Kiland, the commercialisation manager, was particularly concerned about this challenge of production and costs:

So, in terms of costs, there is no risk on the R&D side, because we have the sausage. What relates to money now, is if we can't sell what we produce. If we start a production of three to four tonnes, to test the capacity, what they in reality can deliver, it will cost us a half million NOK if we can't sell it. (Øyvind Kiland, Tine Ingredients)

Thus, here they had a problem, the tension between lack of sales and need for training the production personnel. Kiland tried to balance the issues, although he found that the training on larger quantities was crucial:

First we produced 200 kilograms, and then we produced 500, and this went reasonably well, but we are depending on getting extremely fresh fish within a short timeframe, without brown fat and such things, a quality product. And I said that 'you have to practice producing more than 1000', and then they produced 800, and then I said they had to double, 2000, and then they managed 1700. So they never manage fully, partly due to not getting enough raw material, which is Bremnes' problem becoming our problem. (Øyvind Kiland, Tine Ingredients)

According to Kiland, the reason for not being able to deliver the ordered volumes during this training period was a combination of production skills and supply of raw materials. And the uncertainty regarding the production did not only relate to production capacity (volume), but also the quality of the production:

Two tonnes cost us a few hundred thousand NOK, which is a lot of money, but we don't have a choice. As an example, our latest production now is also the worst. On a scale from one to five, it is a three. The previous production was extremely good, it was a five. I can still bring that one to Japan and Singapore for presentation. (Øyvind Kiland, Tine Ingredients)

The fact that the outcome of *test* production had to be used for *real* customer presentations represented a challenge. Thus, both to achieve volume and quality, the production team had to practise and tune their practise. Kiland was concerned about further training the production organisation, and increased the threat for stocking too much. He chose a balancing strategy, of doing a minimum of training, while he worked on marketing the product:

What can I do with it? Should I enter it as cost of stock, or is it rather just a test production? When it ends up as a level three, it is definitely a test production. But if they should double the production volume again, then we would soon have a stock worth more than a million NOK, and this is what kills projects. We couldn't manage that. So, they have to produce to keep going, with 500 kilogram batches. (Øyvind Kiland, Tine Ingredients)

To build a brand, Kiland argued, you have to deliver the exact same quality every time. This conception was initially not shared by the Bremnes organisation, which also wanted to take the market price for salmon into consideration when deciding on production priorities and type of raw material. Did Kiland go beyond his mandate, by interfering with production? If looking at the formal organisation structure of Salmon Brands, with Morlandstø as managing director, and Kiland representing the sales organisation outsourced to Tine Ingredients, he clearly did. Yet, Kiland experienced his mandate as partly also representing the major owner of Salmon Brands, to do what it took to commercialise the product.

Production competence is both an issue of training and learning for the personnel involved, and at the same time an issue of technology, whether to use hand-crafts or assign and automate the task to a machine. This intersection between economy (cost of labour versus cost of technological investments) and competence (who solves the task best, humans or machines?), was in almost continuous tension, for instance, when Morlandstø, MD of Salmon Brands, considered the task of taking away bones from the pre-rigor fish:

Yes, that's right. The machine that we have developed now helps us a part of the way, but we have to work more on this machine to get it efficient enough. It is also manually driven, we need more camera technology positioning the fish and removing the bones in a more effective way, at the same speed as the rest of the production lines. (Jan Ove Morlandstø, Bremnes Seashore/Salmon Brands)

After the fish had been slaughtered and processed, the product was ready for packaging. There were few places where the relation between production technologies, costs and marketing was more evident. If the product was to be sliced, then slicing technology had to be included, as was the case with questions about whether the package should be transparent, what size the packages should be and what information should be put on them (e.g., tracing info would require information systems, fat content info would require automated infrared measurement, etc.).

Tine did not expect hygiene to become the aspect of production practice in which they had the most to offer Bremnes Seashore. It came to be one of the central aspects of knowledge sharing in the joint venture, indicating the differences also arguably present on sectoral levels, both in different practices of production and marketing, and different competence on micro-biology and industrial nutrients production. Tine's expertise on nutritional standards and micro-biology had been systematically developed during decades of quality work on dairy products, while the Norwegian fish industry had barely been doing product development on fresh products at all. During this process, Tine mobilised some of their production consultants and researchers to teach the organisation at Bremnes how to improve their routines and knowledge. Routines had to be practised and kept sharp on hygiene, regular quality testing and technology maintenance. It should be noted that Bremnes always had been safely within the existing regulations. When, early in the project, I talked to some production workers at Bremnes, they told me that they never had any problems with hygiene; on the contrary, they were very good at this. Still, in the subsequent up-scaling and training phase, they managed to decrease the bacterial concentration significantly. Going closer into the practical work of developing this competence in Salmon Brands, Morlandstø mentioned several issues that were dealt with – for instance, not only how existing routines had to be improved, but also how totally new routines had to be implemented:

> First, we had to drill our own people who are cleaning after the day's work and then we had to drill those cleaning to prepare the next day, and then we had to start up extraordinary routines to check how clean it is. We have registered every single production, and followed up with bacteriological tests that we have been following closely. In some areas, Tine R&D has had to adapt their requirements to realistic levels, while in other areas, we have had to admit that this is much more detailed than we have been used to doing. We have always viewed below 10,000 in total bacteria count[20] as being acceptable, we have typically had a level of 3,000 to 4,000. So we have been very good at this compared to our industrial setting, I think. But when we took this down to 300, or below 1,000, that was a quantum leap, really. (Jan Ove Morlandstø, Bremnes Seashore/Salmon Brands)

In meeting much stricter demands than before, and getting the help to deliver on these demands, they were able to establish a remarkably high standard of nutritional quality, when compared to the industry practice, from this intense collaborative relationship with the agricultural cooperative. Parallel to this process of scaling up production, as the product materialised and the production quality improved, preparations and early attempts at presenting the product to potential users were made.

Looking for users

Where did all these developments and changes lead the project? Did they manage to choose direction related to users and markets, and did they succeed in finding and convincing any of the users they had imagined during product development? I will take a closer look at the process and practice of marketing – that is, of settling a market concept and going out looking for users and markets, and what consequences this had for the project and the product.

Making the concept: Salma

Parallel to scaling up the production at Bremnes during the spring of 2004, plans for international marketing started to materialise, and Hovland was indeed aware of where the main challenge was:

> The main job is done in the market; that is where the sea battle is. I have hired a person to work 100% with development of this as a brand. We will have a test launch with our partners in three markets in June, and then we let them think for a while, and then we take a round in August and September to find out how it is received. (Gunnar Hovland, Tine Ingredients)

Three years after the initial idea, and more than two years after purchasing the patent application, the time had finally come for doing the marketing, in terms of making the product 'marketable', looking for potential users and convincing them to actually buying the product. In having succeeded technically with making a pure salmon salami, and convincing management of the new direction, they earned a degree of freedom in choosing a name and concept associated with salmon that the previous project team did not have. The use of pre-rigor salmon had also solved some aesthetic problems, particularly the fact that the fish salami did not become grey; it had a natural and delicate red colour. Once the white fish was taken off the project and pre-rigor salmon was chosen as the new ally, the conceptualisation gained momentum. The new commercialisation manager, Kiland, started immediately working on the concept and, a couple of months later, he had the concept ready. This meant intensive work on naming, categorisation and designing the product. Several names of a branded product concept had been discussed earlier, partly with help from external marketing consultancy, and names like 'Deli-Fjord' and 'Sea Salami' had been considered to be the best alternatives. However, now they put all previous work aside, and came up with the name and concept of 'Salma', stabilising the association to salmon even more. Plans for sales activities had been delayed some months, as July and August were considered difficult for meeting customers. The production capacity of the new facility was 10 tonnes per month, but they only needed three months to double this volume once, and another three months to double again – reaching a production capacity of 40 tonnes a month – if needed. Kiland expressed the crucial importance of getting sales from January 2005, hopefully from market actors in Japan, Hong Kong, US and France.

During this transition phase between Umi No Kami and Salma, what was left of Umi No Kami's story? As mentioned previously, much of the work to include new actors and resources had pushed the story either into the shadows or out of the project entirely. What was left was, first and foremost, the initial idea about fermenting fish, and the reference to its inventor – Erik Slinde at the Institute of Marine Research. This was the starting point of the story, as was presented. However, the new project constellation argued that their product not was comparable to the patent application any more. The contract between ForInnova, the commercialisation office at University of Bergen, and Tine included points on economic compensation for each new country in which the product was commercialised. For Tine's part, the more they realised how this commercialisation process would progress, the more they realised that these parts of the contract were not very beneficial to them. In their view, it cost more than it gained; thus, they sought to get out of these parts of the contract. They did this mainly by redefining the product, based on the material changes and adaptations that had been done to the original recipe. Among other things, they had gone from using frozen raw material (fish) to using fresh pre-rigor salmon. They had also gone from a mixed product of white and red fish to a pure salmon product. Hence, they argued, they were not responsible on the same contract terms towards ForInnova and Erik Slinde any more. A few other aspects of the history of Umi No Kami were preserved too; the story of a rather messy process at Tine R&D. There was a need for change, for a shift, so the story went. What was not present, however, was Berit Nordvi and her colleagues' roles in the project, how she met Erik Slinde in the beginning, started cooperation with him, and presented the idea within the Tine system. The work on milk proteins in the Neptun project, and the connections between the Neptun and the Umi No Kami projects, was largely absent from the story about what was becoming 'Salma', as it was enacted into the new project organisation. Another historical issue that also was not present, was the cooperation with the Food Research Institute, and the systematic testing of various bacteria cultures (lactic acid cultures), to find what would make the best product in terms of texture, taste and durability in a salami product of fish: how they tested different bacteria cultures from their bio-bank, and how one of them proved to be clearly better than all the others. When I talked to Kiland, Hovland and Swendsen, they consistently explained the product's quality and identity with the raw material – pre-rigor salmon; hyper-fresh salmon of the highest 'sashimi' quality. The bacteria culture, the inner workings of fermentation technology, was forgotten, perhaps simply because it had become stable, working as it should every time, or because they did not know the differences (i.e., the results of the different cultures on the product) or – as hinted at earlier – they needed to create more space for the elements in their new strategy and needed the attention to be on pre-rigor rather than on fermentation and fat stabilisation.

In the final report on the Neptun project, Nordvi made a summary of the relationship between Neptun and Salma. The report is an account of how knowledge produced in the Neptun/Umi No Kami project groups had been used in the Salma product, between the initial intentions and interests of R&D in engaging with this object, and the product that was to be commercialised by Salmon Brands. Seeking to show the project's results and value to its funding and supporting bodies (the Research Council of Norway, Tine, the Food Research Institute and more), the report drew some interesting lines between the stages of the project, shedding some light on what has been a linear development from start to end, and what have been discontinuous technical and political reconfigurations. Here, it is appropriate to ask: to what extent are we talking about the same innovation between Neptun, Umi No Kami and Salma?

While Neptun/Umi No Kami had explored the fermentation of a range of different fish species, both white and red, Salma consisted only of salmon. Neptun had gathered basic knowledge about the curing of meat and fish, and fish quality, and studied opportunities for the transfer of knowledge of cured meat to curing fish. Stabilisation of fatty acids had been a major challenge when curing fish, and in the Neptun project, NIR had been explored to measure the content of fat in the fish, as well as water activity, proteins, and colour in different species. Furthermore, the relation between protein and fatty acid content had been studied, in finding what is possible to stabilise. Salma Cured (the salami), as a pure salmon product, was on the limit of fat content, and succeeded by using only the leanest part of the salmon. Salmon Brands had tested technology for online NIR measurement in their production. While Neptun used only frozen fish, which is common in curing meat because the drying process is then easier, Salma got better stabilising effects by using fresh pre-rigor raw material. The primary aim of Neptun was to study protein applications, which had resulted in knowledge of the effects of various proteins on the fermentation process: pH, loss of water, texture, fat binding, taste, colour, storage stability and, especially, its ability to hinder oxidation (fat fish normally getting harsh quickly). On bacteria cultures, Neptun/Umi No Kami tested several bacteria cultures from the collection of strains at the Food Research Institute, many of which had been isolated from old Norwegian fermented sausages. The chosen bacteria gave a characteristic tasting profile when compared to other salami bacteria, was robust, and had the ability to restrict listeriosis. The bacteria culture was successfully put in production at Tine's own 'filter fermentor'. In addition, a series of parameters for the production process (order of ingredients, temperatures, etc.) and methods of analysis for various purposes (food safety, sensorics, etc.) were developed in the Neptun/Umi No Kami projects and then adapted to the Salma production ('Neptun – Salma' document, January 2005). Technologically, a huge number of alternatives had been explored during the Neptun and Umi No Kami projects. Different fish species, bacteria strains and protein variants had been combined and recombined many times, and several varying technological parameters had been manipulated, such as frozen or fresh raw material, temperature on curing process, measurements of biological composition of the raw materials (NIR) and the order of blending the ingredients. However, here, in the initial period of commercialising the product, of conceptualising a 'brand', its identity and its characteristics were explained solely from the new raw material; pre-rigor processed salmon.

One day, when I came to Kiland's office (mid-June 2004), he was now fully engaged in his new job. He proudly presented the new name of the product: 'Salma'. It had already been decided, and a design agency, Tangram Design, had been assigned to develop a logo and design concept. I got to see a first draft. It was a minimalist logo – 'SALMA' in clean, white letters on black background, with 'Fresh cured salmon' as subtext, and with two symbols above, guaranteeing quality and freshness of the product. The first package design had a bright red colour, picking up the colour of the product. The box, containing a 250 gram whole Salma roll, told a story of the initial marketing preferences, seeking to avoid the extensive work often needed to adapt to various retail actors' demands for sizes and shapes of the product, small or large, sliced or whole roll, etc.

The plans for taking the product into large-scale production (until then it had only been produced in the labs of Food Research Institute and of Tine R&D) were on track and test production was scheduled for the end of the month. At the same time, the plan for launching the product internationally, through food fairs and existing business relations globally, started to materialise. Some of the early choices in the Salma conceptualisation had to do with its category and customer segment, in very general terms. It was to become more of a high-end product, with freshness and quality as sales arguments, than a price-sensitive 'everyday' product. The mundane connotations to salami, at least in Scandinavia, were to be downplayed. Catering actors, such as airlines, baguette producers and restaurant chains, were clearly preferred before retail, due to the relative reduction of adaptation and the assumption that there was more rapid achievement of stable volumes. Nevertheless, retail was still considered to be an option, at least in practice. Finally, Salma was considered to be an international concept, as the Norwegian market was viewed as being probably much too small for such a product. For an object without a category, like Salma, its audience would probably be uncertain about what it was meant to be. When even its inventors were ambiguous about its use, most people who heard about a 'salmon salami', or a 'fermented salmon product', either smiled overbearingly or stated their scepticism right away. Svein Erik Hilsen – chef at Tine R&D, with experience from the Culinary Institute of Norway (GI) – and some of his collaborators let their creativity and craftsmanship loose. The transforming result was striking. Salma was no longer a strange and lonely, and hence confusing, object (see figures in Chapter 3). Salma in combination, Salma with food associates. Thorough combinations of excellent raw materials had been made into dishes fitting different food traditions, hence contextualising the product. Figures showed Salma on display among some of the trendy 'hipsters' of food – sushi and Asian cuisine – hinting at the absolute high-end quality of the product, to be eaten raw, pure and in fine company of other seafood ingredients: seafood pizza, a healthier alternative within the popular Italy-inspired menu; and lefse-roll, a twist on traditional Norwegian food, trading the traditional smoked salmon with Salma in combination with the famous 'lefse' (potato tortilla). Later, these chefs would also appear at various places where Salma needed exclusive introduction to new actors, such as food fairs and business meetings.

Next, Salma started to assemble a set of associations presenting itself as a viable concept for conscious consumers willing to pay to try a healthy and tasty alternative to meat and poultry. In the commercialisation plan for 2005–6, by Kiland (June 2004), the vision for Salmon Brands was to 'set a new standard for Norwegian salmon products, based on quality, uniqueness and freshness', with the goal of creating 'the world's leading brand for product solutions based on high-quality salmon'. Still, were all the 'necessary' associations in place? Would this do the job of selling Salma? It is interesting to note how the discrepancies between interacting practices may create dilemmas and paradoxes. In particular, the difference between the brochures and presentations, which summed up all the best associations that chefs, scientists and marketers could come up with, and the first product version at KaDeWe and Color Line, which were almost stripped of all these associations – opaque package whole roll, no use situations outlined, no visual representations and no opportunity for sensing the product. It was difficult from first impression to tell if the bright red and minimalistic package contained a deodorant or a food product. Slicing would add costs and time for development, and by displaying the product in more transparent packaging, the light could speed up the decay and thus decrease the shelf life of the product.

Looking for users: the marketing tour

In commenting on the early market explorations in Umi No Kami, Kiland revealed a very different approach in comparison with the early Umi No Kami team, which sought out a robust plan for distinct geographical and cultural customer segments. In contrast, Kiland refused to *a priori* narrowing down of the focus for his marketing efforts towards the salmon salami: 'In Umi No Kami they had problems deciding on markets, but we are thinking that all markets are big enough' (Øyvind Kiland, Tine Ingredients).

In being pragmatic about where to sell Salma, he was more concerned about starting to generate sales in practice. Some of the old Umi No Kami participants were critical towards this, viewing it as a lack of focus, while Kiland saw this as a trial-and-error learning process, gradually arriving at a firm market strategy related to the response from potential customers. Still, could he be sure that they really had *something* to sell? That the production had stabilised with sufficient quality and reliability? During a lunch meeting during the summer of 2004, Swensen said that they were on track regarding production:

> Now it is all about market activities, the production is on track. There have been slight problems with differing understandings of quality, mainly small problems regarding hygiene. They are used in sending out products of uneven quality. (Lars Petter Swensen, Tine R&D)

Thus, at this point, Salma was dependent on only marketing according to the team. Having survived ambiguous stages of research and development, and now of scaling up production, the remaining question was whether marketers, distributors and consumers would be able to find common interest in the intersection of economic and use value of the product. As confirmed by Kiland, the quality of the salmon salami now had reached a level where production, taste and concept had become sufficiently stable. We know that they had to work hard on improving production routines for a long time after this lunch meeting at Tine headquarters in Oslo. Another example of this was how Kiland and Swensen demanded that production at Bremnes guaranteed they would produce Salma within a time frame of four hours from slaughter to ready-packaged product:

> We have put a four-hour guarantee on the labels now. They [Bremnes] made it, it helps to be demanding. You know, the salmon industry is still a cowboy industry; it is all about getting stuff out. Bremnes Seashore is better than the rest, but still not terribly good. Lars Petter is continuously following up. (Øyvind Kiland, Tine Ingredients)

In a combined marketing and micro-biological perspective, to be able to write 'four-hour guarantee' on the labels strengthened quality claims, clarifying one of the parameters from which (material) quality and marketing arguments were made: superior raw materials from superior processing knowledge and technology. Salma was now ready to meet customers. Leaving aside the questions of when (how early in the innovation process) to involve customers, and in what way, I will here go through parts of the tour that started in autumn 2004, and continued through the first half of 2005, describing some customers' responses and how the marketing strategy, the product and the concept were changed from these interactions.

The distribution company in the US, Norseland,[21] (a daughter company of Tine), was approached to present Salma in their huge distribution network. They set up meetings with people in the big retail actors, via their established business relations. Yet, would relations based on cheese work for a fish product? Or, in other words, were the same people dealing with both cheese and fish in these giant businesses? The answer was, unsurprisingly, 'no'. The next question was about the extent to which such indirect relations would help. Norseland had little trouble in getting some interesting meetings between Kiland and key actors in the food retail industry in US, and had some useful feedback from them:

> They liked the product, and were interested, although only moderately interested. We will be dependent on a sliced and processed product, but this will demand a lot of work in packaging. And it will be a challenge to maintain the quality after slicing. (Øyvind Kiland, Tine Ingredients)

Again, doubts about selling Salma to retail came up; making a sliced product in transparent packages represented more work than Kiland preferred, although he still took their moderately positive attitude as a confirmation of the product's potential. The American panel also gave 'valuable input' on issues like market segment, size and weight preferences, a need for four rather three months' shelf life, the prohibition of metal clips, etc. Moreover, the response given by the actors from the US on the conceptual side pinpointed some problems:

> Mainly, the U.S. had two concerns: First, they found the price too high, and second, they remarked that the product did not look like what it is, it looked processed – of cuttings – not super fresh. But still, they ate more than just out of duty. (Øyvind Kiland, Tine Ingredients)

This is a good summary of various customer responses, not only in the US. High pricing, from the use of very expensive raw materials and an attempt to signal exclusivity, was perceived as a problem for market entry. Moreover, they presented a product that in many countries 'does not look like what it is', to customers who were more closely associated with mundane and low-price 'industrial salami' than with the more exclusive (and often local) Mediterranean salami specialities. This seemed to add to the problem of price. In a tour going through Moscow before arriving to Hong Kong and Singapore, Kiland hoped that they would have learned enough about presenting Salma to customers in order to ensure the best possible performance in the promising markets of Asia:

> Moscow has lots of 'new-rich' people, but it is a complicated market, it is a 'wild card'. We have to practice, you know, learn what argumentation works, what responses we get. This is important before we go to Hong Kong, by then we have to get rid of all teething problems, and I had hoped to bring our best products there. (Øyvind Kiland, Tine Ingredients)

Before going to Hong Kong and Singapore, a trip with considerable significance, they went to the SIAL food fair in Paris. This time I had the opportunity to participate and observe what was going on, in one of the world's largest food fairs, where big actors meet and cultivate their industrial networks.

I arrived at hall five at SIAL, near the Charles de Gaulle International Airport in Paris, around noon on 17 October 2004. This was an enormous area divided into six halls, which were partially sorted by categories and themes. Eventually, I found Tine's stand, with 'Let Jarlsberg Entertain You' on a large banner. They had only cheese on display, not a trace of Salma anywhere. Kiland and Hovland were busy sitting in a meeting with a potential customer, talking enthusiastically, showing their power-point presentation, and serving Salma followed by white wine around a café table. I soon understood that my opportunity here was to hang around the stand, talk to different people, and see what happened. A representative from Norseland US was optimistic on behalf of Salma on the US market, but still emphasised that this was completely different from the existing portfolio, mostly of cheeses: 'Everything is about relations, Norseland knows people who know people. We can connect them with people who know the fish side, because we don't, it is a completely different market' (Linda Karaffa, Norseland US). In other words, it was unclear how much these existing relations could contribute; what they could do would be more indirect than direct. The outcome of such meetings was uncertain, as was also evident from Kiland's previous trip to the US (see above). The recent visits to the US and Moscow had been interesting, but for the most part as a training exercise for the subsequent presentations, getting the opportunity to test out the argumentation and the presentation materials. Afterwards, the argumentation was adjusted, and new brochures were produced, with better quality and fewer mistakes, before leaving for Paris, and later for Hong Kong to present the concept there. Everywhere, the response for Kiland had been similar: 'The thing is that everyone is positive to the product when they get to taste it.' Initial scepticism turned to openness after tasting the product and hearing the story; but still most actors found it hard to imagine how to sell the product to consumers.

Later at the fair, I spoke to Detlef Martens, Tine's agent in Germany. He was interested in testing Salma on the German market, and reflected on the identity of the Salma concept: 'Perhaps Salma belongs on the meat shelves? It is more similar to those products.' He pinpointed one of the central questions of how to market and sell Salma Cured. How should this new product be categorised? Salma Cured was not similar to anything people had seen before. So, to what category did Salma belong? Or would they have to create a new one, without any established shelves in retail stores, and no associations for consumers to relate to? While Martens here argued for its similarity to meat products, Kiland often argued that the suggested use of Salma corresponded a bit with smoked salmon, used as an appetiser, and in various other settings, often on special occasions. In his view, Salma could also be used in baguettes, or on pizza, etc., which, of course, let him go in the direction of the assumed easier route to successful commercialisation: catering. Further, at SIAL he was mainly directing his efforts towards catering companies. The prospect of getting large and manageable orders from them, less dependent on packaging and similar things, was alluring.

At the end of the first day at SIAL it was time for a party at the Tine stand (see Figure 19). The Norwegian Minister of Agriculture, Lars Sponheim, was there, with his flock of secretaries, journalists and others. Norwegian National Broadcasting (NRK) and a number of Norwegian newspapers and magazines were there, all taking pictures of Holmen (CEO at Tine) and the minister in cheerful dialogue, tasting different tapas dishes prepared by Tine and the Culinary Institute's chefs. Both of them seemed to thrive in the spotlight, and had plenty of time for each other and for the journalists.

Some of Tine's customers and partners were present, as well as Innovation Norway. Generous amounts of wine, aquavit and beer were served, in a pretty cheerful atmosphere. Salma participated in a glamorous fashion, being thoroughly presented as part of several dishes: Salma Cured with dip, Salma Cured alone, fresh salmon (as sushi) alone, and with soy and ginger, in mustard sauce, on canapés, etc. The Salma banner and the brochure had, at this time, been put up for everyone to see (see Figure 20). As with the presentation materials, Salma was mostly

Figures 19 and 20 Salma, the Minister of Agriculture and the CEO of Tine at the SIAL fair in Paris

presented in relation to other things, in recipes of different kinds. Often the interplay between Salma and various other ingredients were emphasised, such as sushi, pizza, baguettes and canapés. By 'teaching' potential customers, using all their skills in communication and presentation techniques, they sought to get the message of *this* salmon product's uniqueness through, as salmon is most often seen as a generic product.

At SIAL, actors from Sweden, Taiwan and Russia, among others, were excited by this new product, and expressed interest in taking care of distribution to their respective countries. Yet none of these really convinced Kiland to make a deal (or was it the other way around?); it was difficult to establish new relations at such a stressful venue, and several of the Tine people, Kiland included, described SIAL as a place where you typically scheduled meetings with your most important customers, as they would all be there. What did convince him, however, was the meeting with certain actors in Singapore a few weeks after.

Catering for Asia

'I haven't heard from them in weeks. We delivered good results on the adaptations of the product; it works excellently on pizza now. I think they just have lost interest.' I had lunch with Øyvind Kiland, commercialisation manager of Salma, catching up on the latest developments of his project of marketing (or rather, making a market for) the 'salmon salami'. He was talking about the multinational restaurant corporation (MRC[22]), with global restaurant brands in their portfolio. There is lots of irony in this story, often found in the instances in which a certain technology, or product, or concept/product portfolio seem stable for some time and to have found their shape, only to be thrown into new uncertainties at the hands of the next actor. After the meeting with MRC, the 'gourmet version' of Salma had to go back to the laboratory. In order to bake well in a pizza oven, it needed less drying, probably no smoking and could possibly accept lower quality standards; in other words, it would be a product that was easier, faster and thus cheaper to make. The process of adjusting and testing a 'pizza version' of Salma in the lab went well. However, when returning to MRC with good news, nothing happened. They had probably lost interest in the product, or they lacked faith in Tine's ability to deliver on their demands, or their contact person had got a new job. Who knows?

The Singapore trip had been timed and planned properly. Coinciding with the official visit of the Norwegian king and queen, and using Innovation Norway as a way of organising meetings and participation in various events, it soon gave results. Two potential customers went into further dialogue about the opportunities for commercialising Salma: Fairprice, a retail actor, and MRC, the restaurant actor mentioned above. The optimal customer to start out with was considered to be an actor within the catering or restaurant industries, which meant reducing adaptations of the product to a minimum. While numerous adaptations would have to be done in relation to retail, a restaurant actor would probably want the product in larger and standardised portions, hopefully minimising technological and logistical adaptations, and so it was the restaurant actor that was the most interesting to the Salma people. Still, they also considered the retail side, together with

Fairprice, a relatively small-sized retail chain, which led to an initial test campaign:

> We got a deal on test sales in Singapore, the 'Fairprice' chain, of both large and small (250 grams) sausages. Had a campaign the two last weekends before Christmas, and we sold some. But it turns out as a niche product, with slicing being a prerequisite. (Øyvind Kiland, Tine Ingredients)

Slicing turned out to be a clear customer demand, and retail remained on the less interesting side of the scale. Nevertheless, slicing opportunities in Singapore and Norway were still investigated, the team obviously intending to go for both retail and restaurant if customer orders materialised.

> It had forced its way through, so that we end up doing retail. And it is like this with new projects, you can't say 'no' either. But of course, in today's convenience market, you have to slice it. But then we have some challenges, one about bacteriology, shelf-life, if we could vacuum it in a tight package, without letting in any light, we would have made it. But there are no customers buying something they can't see. (Øyvind Kiland, Tine Ingredients)

And, where would it be best to do the slicing? Hiring production in Norway would be expensive and, due to the lack of both the technology and the knowledge at Bremnes, it would be as expensive to do it there; and then there was the question of how much the slicing process would reduce its shelf life. It could not be delayed much longer, the pressure for making some sales soon increased:

> I have been pushed hard by Gunnar, and Bremnes, on this thing with slicing, so I feel that it is me who is stopping us, that it is me who is not showing progress. It is really a bit unpleasant, and Gunnar has been bragging a lot about this. (Øyvind Kiland, Tine Ingredients)

In spite of the pressures from his boss and partners, Kiland resisted implementing a retail strategy, feeling that it was too uncertain to defend the work and investments it would take:

> The point is shelf-life, without knowing where your market is, if it is the Norwegian market it is probably ok, except for high costs. But if

you should slice a product, and then send it with a plane overseas, it is really expensive. Or you have to slice it, and then send it by boat, which takes long time, which is bad for shelf-life, hence you will have to freeze it. (Øyvind Kiland, Tine Ingredients)

In addition, the hopes and efforts for making a deal with MRC served to influence this resistance. One of their main R&D units was based in Singapore, and the head of product development showed great interest in exploring this product in pizza recipes. Kiland and Swensen considered this to be an ideal customer, even if having to adapt the product and the packaging here too:

And, [MRC] does want it, and in pre-sliced form. And perhaps they then want a big package, and perhaps they want only 20 slices in each package, corresponding to one pizza. But in any case, they are extremely preoccupied with price, which means that if I add 2 Euros in packaging costs, they will say that 'no, then we'll rather do it here'. (Øyvind Kiland, Tine Ingredients)

These were issues that Kiland wanted to settle before investing in slicing facilities in Norway, or developing retail packages that were both transparent, but at the same time protected the product from light and so forth. On the retail side, no orders did materialise from Fairprice, as the contact person was moved to another position just a few weeks later; Salmon Brands was almost back to square one with this customer. Without contacts, relations, interest, there is not much to do. Geographical and cultural long distance, as well as a lack of knowledge of Tine and Salmon Brands by the new purchasing manager, made it farfetched to hope for more from Fairprice. MRC remained the ultimate customer for Salma at this stage, representing everything they hoped for: restaurants, worldwide distribution and acknowledged brands. It was not strange then, that Kiland became enthusiastic when he got such good initial responses from this actor. On their second trip to Singapore, Kiland and Swensen got to present and discuss the salmon salami with MRC's head of R&D:

They were interested in testing Salma on pizza. But we are uncertain if it is suitable, and in case, it would need less drying, so we are doing tests in our own lab now, on baking, smell, taste and price – as more fat and water will influence price and suitability. (Øyvind Kiland, Tine Ingredients)

The head of product development at MRC suggested that it could be tested in some of their Japanese restaurants as their 'monthly special' campaign in Japan in November or December of the same year (2005), with TV commercials and special offers in the restaurants. This would have meant massive attention to Salma among some of the most open-minded, but also demanding, consumers in a huge market. With more than 200 restaurants in Japan, and normally such campaigns constituting 50–60 per cent of the sales during that period, the potential was huge, even if Salma sold less than normal during such a campaign. MRC had not mentioned any volumes in this round, but a success there alone would probably have made the whole Salma project profitable. Perhaps Salma was now about to find its first large-scale user? As anticipated regarding catering actors, the level of adaptation on packages and logistics turned out to be very sensible:

> A deal with [MRC] would be a great advantage in terms of packaging and labour, they wanted Salma ready sliced in packages of 1–2 kilograms, without any demands on its look, and then vacuum packed and frozen. Salma will then have three weeks' shelf life after defrosting, while pepperoni will only have one week. (Øyvind Kiland, Tine Ingredients)

However, this customer also had its labour-intensive demands, and to be able to answer the question of feasibility for warm food properly, Salma had to be taken back into Tine's laboratory. From finally being stable both in shape and production, its identity was now questioned again, opened up. Would it work as a 'pepperoni-substitute' on pizza? A short but intense period of experimenting and adapting to MRC's requirements produced results:

> There is a new variant of Salma, less dried, containing more water, but besides that using the same technologies and recipes. The time in the drying facilities was reduced from 3 to 1 week, hence tripling the production capacity. The result is a product well-suited for pizza, because it will not curl from baking, and it will not become as dry in the edges as pepperoni. (Øyvind Kiland, Tine Ingredients)

The fact that Swensen joined Kiland on his visit to this customer was emphasised as an important part of establishing the good process, enabling them to answer technical questions on the spot, and giving a more fluent professional dialogue on relevant aspects both of product and market, in addition to providing Swensen with better understanding of the customer's needs. Moreover, they saw this work on developing the product with one of the 'big players' as a great opportunity that could be exploited in relation to other customers:

> I think we will use this relation to get access to other places, too, having developed a product that is well-suited [for pizza], and that we can refer to an interested actor. For example the frozen pizza market. (Øyvind Kiland, Tine Ingredients)

After a process of altering some of the steps of processing, such as salt, water content, smoking, testing times of storage and finding the best solutions for slicing and distribution, the results were positive, and they prepared for negotiating a deal. However, then problems arose:

> All [MRC]'s demands have been met, and we will deliver 1 kilogramme packages, frozen and sliced. But [MRC] is not answering. Lars Petter came with me, talked with their researchers and product developers, and we have developed the perfect substitute for pepperoni. I don't know why they have cut the contact. (Øyvind Kiland, Tine Ingredients)

A new variant suitable for warm food had now been created, 'the perfect substitute for pepperoni', and with even better properties on storage time and behaviour in the oven. Despite these first-class results, this part of the Salma story also ended before it reached any users, in this case, restaurant guests in Japan, Singapore or Hong Kong. When the good results were sent from Tine to the MRC R&D headquarters in Singapore, nothing happened. Silence. The process of adjusting and testing a 'pizza version' of Salma in the lab went well. Notwithstanding, the customer had, for unknown reasons,[23] lost interest, and the attempt to mobilise the desired customer had brought much work, but then failure.

Let's recap this story from another angle. From excerpts of the email correspondence between Tine/Salmon Brands and MRC during 2005, we can recall parts of the interaction of this attempt at establishing a customer relationship:

> 2005-01-10, email from head of product development, MRC R&D, to Hovland/Tine:

> We have generated a lot of interest in Salma amongst our franchise in Asia. Apart from Japan, both Hong Kong and Taiwan would like to

further explore this product. Can you please send a log of the Salma with the bigger diameter to Hong Kong and Taiwan? Sorry for the short notice but your help in this matter is highly appreciated.

The first presentation of Salma had obviously made some impression on the MRC organisation, which was very interested in exploring this further. They had been discussing how Salma could be *used* in MRC's context (on pizza), how Salma then would need to be modified, and how to market it. Now they wanted to test the product, both in their own laboratory, and with selected regional managers. Concurrently, Swensen worked with the required changes at the Tine R&D lab to get it adapted to pizza production. After some weeks of corresponding, talking and developing, based on the feedback from MRC's representatives, Kiland could come back to MRC with good results on their requests:

2005-02-11, email from Kiland/Tine to head of product development in MRC R&D:

Our R&D department has been able to adjust both the production method and the recipe in order to make Salma more suitable for pizza. In addition we 've been able to cut costs with approx.20%.

Results are promising:

- There is no evaporation of fat at 250 degrees for 8 minutes.
- Neutral taste.
- Spice upon request can be added.
- Unique process (difficult to copy)

On the other hand, this product needs to be kept frozen during storage.

Solving the technical changes and at the same time reaping economic gains by reducing costs was clearly an uplifting message to send to Singapore, and MRC responded immediately:

2005-02-11, email from MRC R&D to Kiland/Tine:

Thank for your follow up on this project. As for the samples, I would prefer evaluating them first before sending to any [MRC] Markets. Lastly, is there any update on the slicing of the Salma?

In retrospect, things could be seen to start falling apart already at this point. Anyway, two weeks later Kiland got impatient, and sent a reminder to MRC, using the production schedule as a means for speeding up and getting some feedback:

2005-02-25, email from Kiland/Tine to MRC R&D:

We 're anxious to hear your comments regarding our latest production. Our R&D department are planning a new production on Tuesday next week. In order for us to take your inputs into consideration, we'd very much like to hear from you.

A few days later an answer came, still showing their interest and willingness to take the development process further:

2005-03-01, email from MRC R&D to Kiland/Tine:

We tested Salma #4 - Natural.
  Generally, it was a great product but just needs a little fine tuning.

  A) Some improvements we would like to have are as follows:
       1) reduce the salt level slightly, 2) increase the moisture
  B) In terms of the Pastrami and Pepperoni flavoured samples, how different are the costs compared to the natural?
  C) Is there any further opportunity to work on reducing the cost?
  D) We are also interested to know on the possibility of having the product pre-sliced at 1–1.5mm thickness. Any progress on this?

Lastly, if you could work on the improvements today, would appreciate if you can send us the improved samples soonest possible.

This was enough for Salmon Brands to keep on working for the deal, a new production of Salma, adjusted to meet MRC's suggestions was made, and answers to MRC's questions on recipe, slicing and economy were provided:

2005-03-03, email from Kiland/Tine to MRC R&D (excerpts):

The production this week went according to plans, and we 've managed to reduce the salt and increase the moisture.

Products will be ready next week, and we will of course ship it to you as soon as possible. Most likely on Thursday 10th.

Here is my reply in regards to your questions listed below:

...

B) Flavour adding does not change the price of the product....

C) By increasing the moisture, the product automatically gets cheaper in production. We've reduced cost by approx.20–25%. I 'll need to get back to you on details as soon as we've done the calculations on the latest production.

D) The sliceability of the product is very good. The question is where to slice it, in order to maximise shelf-life, quality and optimising costs.

Thus, after sorting out every question, adjusting for every technical demand and another round of estimating the (reduced) costs of Salma, they just had to wait for a decision from MRC if they wanted to take the next step and bring Salma to the market test. But quite a while passed without hearing anything more from MRC, and as late as June, another reminder was sent:

2005-07-06, email from Kiland/Tine to MRC R&D:

After our last visit in Singapore, we've been developing our product further and always tried to keep the needs of [MRC] in mind. We now have the processing plant in place, which includes slicing and packaging. Please let me know when and where you 'd like us to send samples.

No response. If not before, now hope was definitely fading. MRC had, for unknown reason, lost interest. Another reminder was sent, just to make sure:

2005-08-24, email from Kiland/Tine to MRC R&D:

I'd like to remind you that our Salma Pizza Roll is now available for potential buyers. It has all the way been in our interest to serve [MRC] first. Hope to hear from you.

The efforts seemed partly wasted, no deal this time either. The prospect of the ultimate customer had met its end, at least temporarily, and the demands for making sales kept growing. This was not, however, Tine's first experience with demanding Asian customers. With regard to cheese and also on biomarine ingredients (Maritex) they had encountered a very different approach to business relationships from that which they were used to in Norway and Europe, especially in relation to the level of resources pulled into the relationship, and the intensity of interaction. Hanne Refsholt, CEO at Tine, commented on the challenge:

> One thing is being thorough and accurate, but it is coming up new topics on the agenda all the time, making this into a continuously ongoing process. Hence, when you have innovation processes that are to be customer driven and involving, when you in addition go abroad, then you have both different industries and also many different cultures, which add in some aspects of uncertainty, and extra needs for competence and the ability to drive things forwards and produce results fast enough.

Some lessons had been learned in these relations, as when Hovland and Kiland brought the R&D competence of Swensen with them to Singapore and other places, understanding how the various professions talk with their own 'tribe' easier than across professions, especially when the cultural differences were so big. Yet this time it could not take them far enough, and so the 'industrial track' of Salma was largely dead. The last hope for the salmon salami at this point was its only real end-consumer test, the launching of 'Salma Lax Salami' in German hypermarkets.

A market test for salmon salami in Germany

After Tine's agent for distribution of cheese in Germany, Detlef Martens, showed an interest in Salma at the SIAL fair in Paris, plans for distribution to retail chains in Germany started emerging. Around the same time, also Color Line, a ferry company with a route between Kiel and Oslo, agreed to sell Salma in their on-board shop. Interestingly, it sold well there, in particular to German tourists wanting to bring something 'Norwegian' back home from their Norwegian vacation. Hence, there was some hope that Germans would be interested in Salma in their own food stores too:

> [Martens] has been Tine's man in Germany for a long time. Dynamic and clever guy, has many contacts, and faith in the product, really wants it, so now he will get it. Perhaps we will sell 50 tonnes the first

year in Germany, in my head I think 10, but if he can make 50, he has saved the project. (Øyvind Kiland, Tine Ingredients)

In a way, Salma was not 'sold in' to Martens and the German market, it was Martens asking for it, and hence the responsibility and drive for making this happen was on Martens' German organisation. They soon had the initial, unsliced, package for sale in KaDeWe, Berlin's huge and prestigious demonstration store for food products:

> We had to have knowledge for our presentations, about what it is, how it is made, and what is in it? So, when taking it to our customers, we had to explain it to them. We also had to find a way to lower the price, because 10 Euros for a 250 gram package is a very high price. We tested it in KaDeWe in Berlin, and we had promotion women to present it and give out tasters, and in a few days we sold 10 cartons, 100 salamis. (Detlef Martens, DM-Nor)

Several lessons were learned in this preliminary market test. First, customers needed knowledge about the product – what it was, how to use it, its benefits compared to alternatives, etc. This had to be inscribed on the packages and presentation materials, as well as having 'promotion women' in the stores for presenting and giving out tasters of the product. Second, slicing and decreasing the size would clearly be beneficial. The package sold in KaDeWe was not sliced, it was not transparent, and its size (250 grams) made the price too high. It was the second preference, retail, the assumed work-consuming alternative, which was once more activated, but with an important difference from the Fairprice case. There was already an established relationship in place, on cheese, which in this case study turned out to be useful also on fish, getting access to a number of large hypermarkets in central regions around Berlin and Hamburg:

> There are very few brands on salmon. We will not do general marketing, but rather build a brand on the German market. Fish is mostly competing on raw material. Our relationships with these chains make it possible to enter the fish area; I have been in the trade for 20 years. [But we still] have a job to do in convincing the buyers. (Detlef Martens, DM-Nor)

A novel thing about Salma was the development of a distinct brand, which was very different from the generic marketing of salmon in most places. In preparing for presentation to the relevant retail chains, the design and packaging for the German market was developed, and sizes were decided: small packages (50 grams) to reduce unit prices, ending up with a price of approximately 2.50 euro to end consumers. Armoured with presentation materials, a suitable package and a novel and branded product, Martens could go to the retail chains with the product:

> When we presented it to them, they were initially astonished; it was brand new, they were positive. Then, they calculated the packages and the prices, and responded that it was very expensive. (Detlef Martens, DM-Nor)

In being presented arguments for why consumers would be willing to pay for such a product, launching it as 'something extra for the weekend', purchasing managers were convinced and willing to give it a try. Salma was ready for test sales in 90 German 'hypermarkets'. But would consumers be convinced to try out a 'lax salami'? How should it be positioned and communicated?

> It is very different from meat, but to be able to teach people about it, it is best to describe it as a salami, but that it is meatless, lower in fat, very healthy. And then people become very positive to it. It is definitely an alternative to meat. We will present it in what we call 'fine-cost shelves', which is where we normally sell things like caviar, smoked salmon and things like that. (Detlef Martens, DM-Nor)

Both similarities and differences from meat were considered important. Meat salami was helpful for finding associations to category and use, but the nutritional value and the supreme raw materials had more associations to fish products; hence, there were attempts to make certain associations to both by calling it 'salami', although placing it among seafood products in the store's shelves. Although not catastrophic, the sales of the 'Lax Salami' did not go particularly well, not even after adjusting the packaging information and trying a second round. With the specific marketing challenges of such a category-crossing product – such as having to run demo-stands in the store in order to convince people to buy it, not knowing how best to position it, and not having budgets for mass-media commercials – it was soon realised that Salma, in this form, had little chance of commercial success, even in German hypermarkets. Thus, one by one, Salma's potential partners and associates for marketing and distribution either did not connect at all, or failed to fulfil the promises of the partnership. The lack of commercial results, of course, disappointed the owners at Bremnes Seashore, and the relationship between Kiland as commercialisation manager, and the production management at Bremnes started to disintegrate:

> We imagined this to become a good project, a big project, that would give us stable prices on our salmon, and that the volumes would be larger. But it hasn't. It has been a big disappointment, nothing else to say about that. (Olav Svendsen Jr, Bremnes Seashore)

Clearly disappointed by the lack of commercial progress, the owners at Bremnes gave a rough critique of the marketing and sales efforts of Tine Ingredients. A main reason for partnering with Tine had been the hope to get better prices on high volumes of their fish. Yet, it turned out, it was not only the sales performance that was problematic, it had also to do with disputes on technical developments, where Morlandstø and Kiland had both communicative problems and differing views on necessary technology:

> Between me and Øyvind [Kiland], things broke down after a while. I think that when Øyvind came into the first phase in our collaboration, he came to think of himself as managing director in Salmon Brands, that he in a way bought the services he wanted here at Bremnes. (Jan Ove Morlandstø, Bremnes Seashore/Salmon Brands)

This breakdown of the personal relationship between Morlandstø and Kiland, and Svendsen Jr's disappointment in Tine not managing to sell the product, serves as an illustrative end point for the Salma Cured story. The story of salami of fish was fading out, and the involved actors slowly had to acknowledge that this would not become a commercial success, at least not in the short-term. Gradually, focus among the constellation of actors around Salma shifted from selling the salami ('Salma Cured'), towards making a fresh loin version, called 'Salma Fresh'.

Fish salami unmade

This story could have been told from another angle entirely. It could have been framed as the invention of new processing technology for farmed salmon,[24] – what is called 'live cooling' and 'pre-rigor processing'; the scientifically documented techniques that provide considerably higher quality on fresh salmon fillets. This story line started with

the small fish farm and processing plant, Bremnes Seashore, on the
south-west coast of Norway, and continued with a set of collaborative
research projects between Bremnes Seashore and a research group at
the Norwegian University of Life Sciences. Along this story line, the key
problem would have appeared when the people at Bremnes – in succeeding in making supreme salmon loins – discovered that they could
not get extra value from this new product in their market, or we could
say from their distributors. Yet having spent 40 million NOK on the
matter, Bremnes Seashore was of course very keen, perhaps even desperate, to find ways to get returns on their investments. So, when their university researchers, now with jobs in one of the largest food producing
companies in the country, came back and asked if they were interested
in a new project – on fish salami – they were not hard to convince.
This is the point where the pre-rigor story crosses the fish salami story,
and where much of the present case study is situated. However, what
became evident in the aftermath of the salmon salami project was that,
in the end, pre-rigor took over the whole constellation. When Hovland,
Kiland, Swensen and the others at Tine succeeded in transforming
Salma from a product brand for a salmon salami to a portfolio brand for
fresh salmon products, the road was open to selling fresh loins instead
of the salami. I have not seen any formal decision putting the salami on
a sidetrack, but when 'Salma Fresh' saw daylight, soon after the failures
of Salma Cured in Asia and Germany, the fresh loins soon got all of the
focus of the participants. One of the responses from the US retail directors (see above) was illustrative for the emerging understanding in the
project team that it would be much easier to sell the raw material than
the salami:

> When I did my presentation, first of the pre-rigor salmon, and then of
> the processing of a salmon salami, one of them responded, in broad
> American dialect: 'Never mind about the sausage, where can I get the
> fish?' I could have signed a deal on fresh pre-rigor salmon right away.
> In a sense, we have the freedom to continue as we please, but the sau-
> sage is still a precondition. (Øyvind Kiland, Tine Ingredients)

If Kiland and Salmon Brands had not been convinced about including pre-rigor fillets as part of the Salma concept before, this crystal clear
statement from the American certainly provoked some action. In both
giving Kiland a stronger conviction, and circulating around in Bremnes
and Tine for months afterwards, the event became a powerful story
explaining why Salma needed expansion. At that point, they could not

cut off the salami, as it was still the reason for doing this at all, but it contributed to a shift of mind, speeding up the inclusion of fresh fish as part of the evolving strategy. This emergent idea of including fresh loins also came up in some of my conversations with Kiland and others at the SIAL fair in Paris:

> I am having a lot of discussions with the Americans. It is fun, they are really pushing. We will try to sell in fresh fillets in Costco, a large retail chain, that today buy post-rigor fillets from Norway. We have to find out who we compete with, as we will have a different pricing. Very few actors are inside with Costco, but they sell big volumes. (Øyvind Kiland, Tine Ingredients)

When talking to Kiland at the fair, it struck me how he was clearly drifting towards a change of strategy, being of the opinion that, in many settings, it would have been easier to sell super-fresh fillets first, and then sell the sausage. The idea that had been more or less present from the 'coup' had transferred the project to Tine Ingredients and Salmon Brands, and strengthened from the first meeting in the US. Also at SIAL, he was continuously making sense of the project through meetings with customers, colleagues and partners. Sometimes he was almost doubtful about the potential for selling the sausage at all: 'But think about it, would the sausage have been the product you would have started with?', and at other times negotiating and doing sales presentations of the sausage to various actors from different countries. He had kept, and strengthened, this emerging strategy in his customer presentations (power-points), talking first about the superior pre-rigor salmon fillets as a prerequisite for the salami, and thereafter presenting the salami. He felt that this worked better.

The full story of this transition and the subsequent quite successful marketing of Salma Fresh will not be told in this book. To be very brief, the story – from a Tine perspective – is about an entire turnaround of the market strategy (again). The fresh loin, Salma Fresh, was first presented at local fishmongers in Oslo, but they were not at all interested in selling vacuum-packed fish loins. This packaging design, with emphasis on the Salma Fresh brand, soon after got a reward for good design from the Norwegian Design Council. Kiland then brought his delicate packages of salmon loins to a high-end supermarket, Jakob's, and their fresh produce manager Roar Sjåvåg immediately caught interest. Within a couple of weeks they were ready for an introduction campaign in the store, which gave great sales at good prices for the new product. Jakob's was associated with Norway's largest retail chain, NorgesGruppen, which then became very interested. It must also be said that Tine, as the dominant supplier of dairy products for almost a century, had close relationships to this retail corporation. Soon a roll-out plan was launched, first for launching Salma Fresh in another two supermarkets, Centra, before getting it out in a larger number of their 'Ultra' and 'Meny' supermarkets. At the same time, several gourmet chefs had found this to be a great raw material for their cooking, which ended up not only helping Tine with making marketing materials, but also putting Salma Fresh on their menus and serving as charismatic ambassadors for Salma worldwide. As this is being written, Salma is swimming rapidly towards a great commercial success story. This is also the public version of the story, as it is told by its marketing representatives and by the press. The salami is never mentioned as part of the concept any more, at least not in public. The crooked path towards this success story, whether on the Bremnes or on the Tine side, is carefully deleted.

Finally, this story could also have been told as a series of discontinuities.[25] First, it is a story of Tine buying the patent and starting Umi No Kami instead of supplying another producer of fish salami with whey proteins. Second, it is a story of not having the patent application approved, and departing quite a bit from the original recipe to manage the product, in the end producing from fresh salmon instead of frozen saithe and salmon. Last, it is a story of departing from fermentation altogether, instead commercialising the raw material as Salma Fresh. In this last transition, the Neptun final report was right and wrong (see above). A prototype three was indeed developed, Salma Fresh, but it did not use the knowledge or technology of whey proteins and bacteria cultures in the recipe. In this sense, Bremnes' pre-rigor salmon became both the saviour and betrayer of the fish salami. The project was continued with new vision and energy with the new ally, but in the end the raw material took over the project, putting Salma Cured on a sidetrack indefinitely.

Part III

5
An Analytic Scheme of Innovation Processes

Given that my research questions are 'how questions' – how do innovation processes evolve? – Chapter 4 has basically answered the main question. It is a relatively detailed description, based partly on real time and partly on historical accounts of an innovation process. It reveals many of the uncertainties and contingencies of innovation processes that, in most academics' and practitioners' accounts of innovation, are black-boxed by lack of real-time observation, oblivion and post-hoc rationalisation. However, we still need to attempt to clarify what we could learn from this. By taking the case study observations and combining them with process-based analytical and methodological thinking, I will do two things. In this chapter, I will outline an analytic scheme (or a 'conceptual model') for studying and analysing (industrial) innovation processes that seem better suited for investigating *the practice of industrial innovation* that can be found in the established repertoire of actor-network theory, as well as in some recent contributions to innovation management. The framework I suggest here did not come *before* the fieldwork; rather, it is an outcome of the combination of process-based theory and my observations in the field, of recording and trying to discriminate between what kind of activities and 'sub-processes' are happening in practice, and pairing that with the logic of a relationalist and process view (Latour, 1987; Hernes, 2007). Second, in Chapter 6, I will draw out the theoretical implications from reconstructing the case study via the analytic scheme, suggesting how my study may contribute to, and complement, theories of innovation processes.

Based on my fieldwork, and the methodological-analytical basis provided in the first chapter, I suggest that innovation processes may fruitfully be conceptualised as a dual process: first, as a process of *mobilising actor-networks*, or getting the rights, alliances, space, time and resources to innovate; and, second, of *knowledge exploration* in formulating and testing propositions about reality, which also means interacting, because reality (people and things) often 'speaks back' (see Figure 21). Thus, we get a bipolar model, in which the particular dynamics between the two poles of a concrete innovation process become a central part of explaining that case. This explanation, in turn, may have more or less relevance for understanding other innovation processes – that is, for theorising (see Chapter 6). This dual process happens within a *network of interconnected processes*, which create resistances and constraints, as well as enablers for innovation in certain directions rather than others (see Figure 22). Moreover, the initiation of innovation processes needs some kind of 'staging', or bringing of attention to ideas from boundary spanning actors and putting an innovation process in motion. Last, it is clear from the case study that the processes of mobilisation and exploration are neither completely separate, nor completely intertwined. How and when these sub-processes interact, and the implication of this, seem to be important questions for understanding innovation processes.

Still, can a productive distinction be made between these two processes of mobilising actor-networks and knowledge exploration? Is not

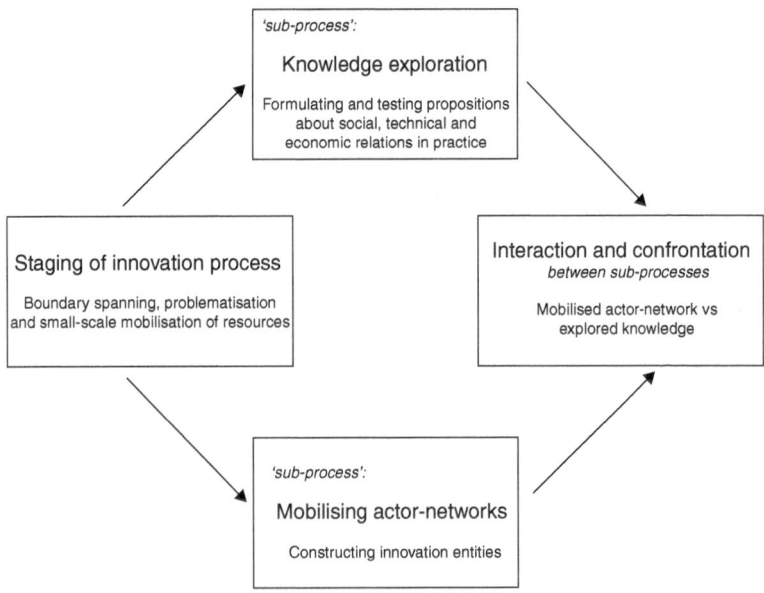

Figure 21   An interactive process model

knowledge exploration also a matter of negotiating socio-material relations (as in mobilisation)? What then is the difference? Several of these activities, although analytically separable, are in reality interacting with each other. Mobilisation activities may have more or less immediate influence on the exploration, and vice versa. However, it takes different tools – skills, strategies and resources – to (1) recruit and mobilise elements, and (2) make them fit and hold together. The first, or 'mobilising actor-networks', typically consists of political activities/processes of (re-)presenting, convincing, forcing and negotiating. The second, or 'knowledge exploration', consists of knowledge generation. This involves the exploring and stabilising of relations between elements, such as ideas, materials, technologies and procedures; formulating propositions and testing them in practice. As will be clarified in Chapter 6, there are also differences in outcomes: exploration/knowledge generation tends to increase uncertainty, while mobilisation/power tends (if successful) to reduce uncertainty. Paradoxically, as the exploration process tends to produce multiplicity, ambiguity and thus choice, it demands, at some point, interaction with the mobilisation process. Choice under uncertainty needs vision, ideas and/or programs to mobilise (power for) settlement of controversy (Latour, 1988).

When studying innovation processes from this perspective, it is not enough to just trace these two kinds of processes as if they were separate. A critical issue is when and how these processes interact, and what kind of tensions and/or opportunities such interaction produces for the innovation to realise and find use, as stated by Latour: 'In order to follow a technological project, we have to follow simultaneously both the narrative program and the degree of "realization" of each of the actions' (Latour, 1996: 81).

The distinction between mobilisation of actor-networks and exploration of knowledge has a parallel in Latour's distinction between the 'narrative programme' and the 'degree of realisation of each of the actions' in technical innovation, where there is rarely a one-to-one relationship between them, and where they are not always very closely connected, but time and again have to interact, often causing confrontations, conflicts and needs for negotiations and compromises. The tension between the 'promises' of the narrative programme and the 'realities' of the realisation process is not settled by achieving consistency between them, but by coming up with a reality – in one way or another – that serves the interests involved.

Before explaining the analytic scheme in some more detail, I have to make certain reservations. When I use the terms 'process' and 'subprocess', I do not mean to say that there is an objective 'whole' that may be divided in distinctive parts. This is, as with most conceptual models and analytic frameworks, rather a matter of delimiting cases, research objects and research questions. If required by my research interests, I could have zoomed in (e.g., only covering the change in production routines) or out (e.g., looking primarily at Tine's biomarine strategy and portfolio), and hence delimited my case and research objectives differently, and still been able to use the same analytic scheme. What I call sub-processes in this case study may well have been parts of other processes too, serving other interests, and in fact this is part of my argument as explained in this chapter and the next. The interconnecting (or embedding) of processes is a crucial challenge during innovation processes. There are numerous examples of process-oriented researchers that pragmatically develop conceptual frameworks that emphasise certain parts of the process, such as Law's (1994) 'modes of ordering', Callon's (1986) scheme of translation, Latour's (1999b: 100) model of circulation of scientific facts and Kjellberg and Helgesson's (2007b) translation model of market practices. Not dissimilarly to these authors, my aim has been to (pragmatically) emphasise parts of the innovation process that seem particularly important in my case study. The model is depicting aspects of innovation *processes*. It does not refer to entities or actors but rather kinds of activities and processes and their logics. The reason why I have developed my own analytic scheme instead of using existing models is discussed and demonstrated in Chapter 6.

Staging of innovation processes

The initiation of innovation processes is what I call 'staging'. Something happens somewhere; someone asks a question, problematises something or incidentally discovers something. In my case study, we saw how new technoscientific ideas emerged from curious experts (technologists, scientists, managers, etc.) spanning the boundaries of their knowledge, facilitating interaction between actors and elements from different epistemic and/or industrial fields. An idea comes out of the meeting between different perspectives, different realities, different knowledge and experience. In industrial settings, we may call these entrepreneurial actors 'boundary-spanning experts', using their curiosity to explore the boundaries between different fields of knowledge. Often it is technologists seeking to supplement technologies for recombination or seeking potential use for their inventions. At other times it is marketers or customers that formulate a demand for a solution to a problem, or managers that seek to renew their organisation. What all such situations have in common for starting a process of realisation is that the idea has to be brought to the attention; creating interests and mobilising a minimum of time, space and resources. In the case study of Umi No Kami/Salma, problematising questions of the future of industries (competition in agriculture, lack of industrialisation in aquaculture) together with a high level of technical fascination both helped the formulation of the idea and the creation of attention and interest during the early stage. Moreover, it was fronted by a senior professor with a wide professional network who was employed at a research institute – that is, he was simply doing what a scientist should do: formulate and investigate new questions. Sometimes it is necessary to stick with the initial question for a while before finding an opportunity to do something with it – that is, stage a process of exploration and mobilise resources (Spinosa et al., 1997). To cultivate the ability to formulate questions beyond the present knowledge domain and industrial path takes capacity for 'mindful deviation' (Garud and Karnøe, 2001).

Mobilising actor-networks

After formulating the question, putting it on the stage and creating some interest in it, the problem immediately arises of how to mobilise the time, space, actors and resources needed to start the exploration and realisation of the innovation. Slinde partly had to show that there was something to explore there, through initial experiments and convincing presentations of the premature materialisation of the idea. Further, he had to enrol actors with money and expertise to participate in the further development, in this case by exploiting established relationships to people within funding bodies, a technology transfer office and at his present and previous research institutes. However, this was not a one-time operation. Throughout the process(es) described in this case study, the innovators repeatedly had to mobilise renewed support and more resources from their allies, and/or to find new partners. We also saw how a number of arguments were used, and how several actors were appointed as representatives for the project – or rather for the *potential* of the project. This is a pragmatic process; making the most of what you have when using it in convincing and negotiating to expand the actor-network and the access to resources. More or less coherent *chains of arguments* have to be constructed to produce mobilising power over resources and decisions. In addition, there is the work of enrolling and aligning a set of actors and resources into an 'actor-network', making them represent and support the project, and keeping them interested over time, while doing the exploration work. During exploration, the immature object has to be taken through several translations; from idea, to prototype, to research application, to patent application, to product, to use and exchange, etc. This partly depends on mobilisation; the construction of meaning and the mobilisation of chains of arguments: a technical recombination is possible – someone might be interested in using it in his/her practice – thus it could be exchanged for money. Exploration, on the other hand, is about testing whether these ideas and propositions hold in reality. Is (or can we make) this technically feasible? Does it (or can we make it) fit within the distributor's product categories? Does it (or can we make it) fit within the using practices of consumers? How much are they willing to (or can we make them) pay for it?

Knowledge exploration

The other side of the coin is that when you actually have succeeded in mobilising some resources and convincing some people to give the idea a try, then you have to make it work in practice. This process of knowledge exploration,[1] of 'making things work', involves a process of formulating and reformulating propositions about the (potential) 'reality' of the innovation, and then testing it out in practice. It is a two-step process: first, of creatively imagining potential social and technical relations, and then testing in practice if – and in what way – such relations are possible. An analogy to this process would be that of the scientific method, of formulating a research question, or a hypothesis, and doing practical empirical experiments to see if the answer to the question can be found, or if the hypothesis may stand the test. This is not a one-way street of an actor seeking to impose his/her will on another, but rather an interactive or, we could say, negotiated process. When testing a relationship between elements – for example, between fish and fermentation culture, or between salmon salami and users – the innovator enters her own picture, so to speak, and becomes involved with – and a part of – the object. Thus, not only is the innovator testing a relationship between elements, but the innovator herself experiences how the elements 'speak back' – that is, accepting some relations while rejecting others. Moreover, this testing of – and making of – relationships changes the innovation, often in unpredictable ways. Thus, although equipping the innovator with some creativity and agency, innovation as recombination is not a matter of unlimited agency of the heroic entrepreneur. This is not a relativistic 'anything goes' perspective, as long as the political work of mobilisation of actor-networks succeeds. The proposed relationships between technical, social and economic elements have to be tested and negotiated, and then reformulated and renegotiated, often several times. An unavoidable aspect of exploration is uncertainty. Precisely because exploration starts out with imagination, and because the objects being explored 'speak back' and bring in their own preferences (enabling some relations while resisting others), it is not possible to be sure whether – or how – the imagined recombination of elements could work. Knowledge exploration produces development risk, as there always will be 'nobody-knows' problems present, and in the case of more radical recombinations, the number of such uncertainties causes an almost indefinite development risk. This part of the 'translation' process of knowledge exploration is about developing 'chains of propositions' – from testing whether a technology is feasible (including all its supporting technologies), to testing whether such a product will find (paying) users and hence produce economic value for the innovating actor-network.

Interaction and confrontation of sub-processes

However, as mentioned previously, processes of mobilisation of actor-networks and knowledge exploration are not fully separable. Sometimes (but definitely not always) 'chains of power/arguments' (mobilisation) and 'chains of propositions' (exploration) interact with each other; borrowing elements from each other, or confronting each other's aims and outcomes. This does not continuously happen. After mobilising a set of actors and resources, and starting the exploration for shorter or longer periods, there may be no or little interaction with the mobilisation process and the original idea and intention. Parts of it may be involved in formulating and testing various propositions, while other parts may not. Yet, then, when resources run out and new resources have to be mobilised, for example, or when allies in the actor-network start getting impatient or are disappointed, the exploration process may be confronted for its lack of progress, its departure from the original vision or its need for reorienting towards enrolling other and different actors and resources. Similarly, discoveries and knowledge generated in the exploration process may challenge the mobilised actor-network to rethink and change their ideas, interests and participation, both allowing and supporting the project in exploring new directions and propositions. Both the ways in which such interactions and confrontations come about and what they lead to should be of particular interest, as this

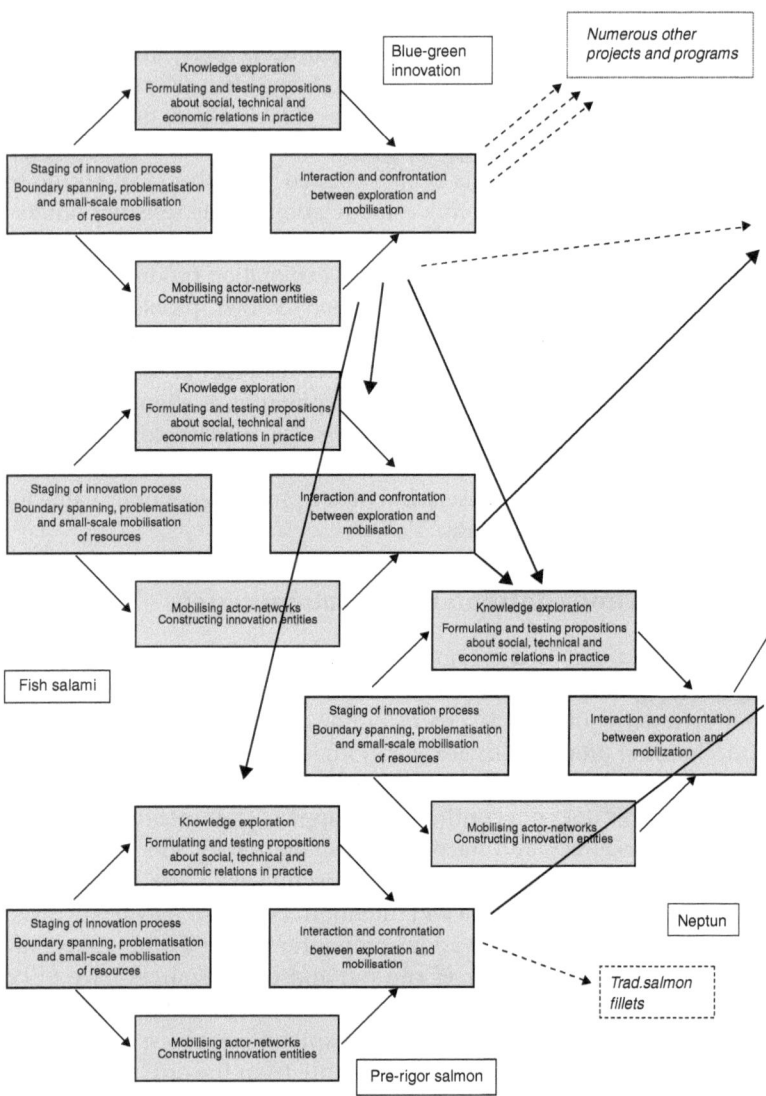

Figure 22   Networks of interconnected processes

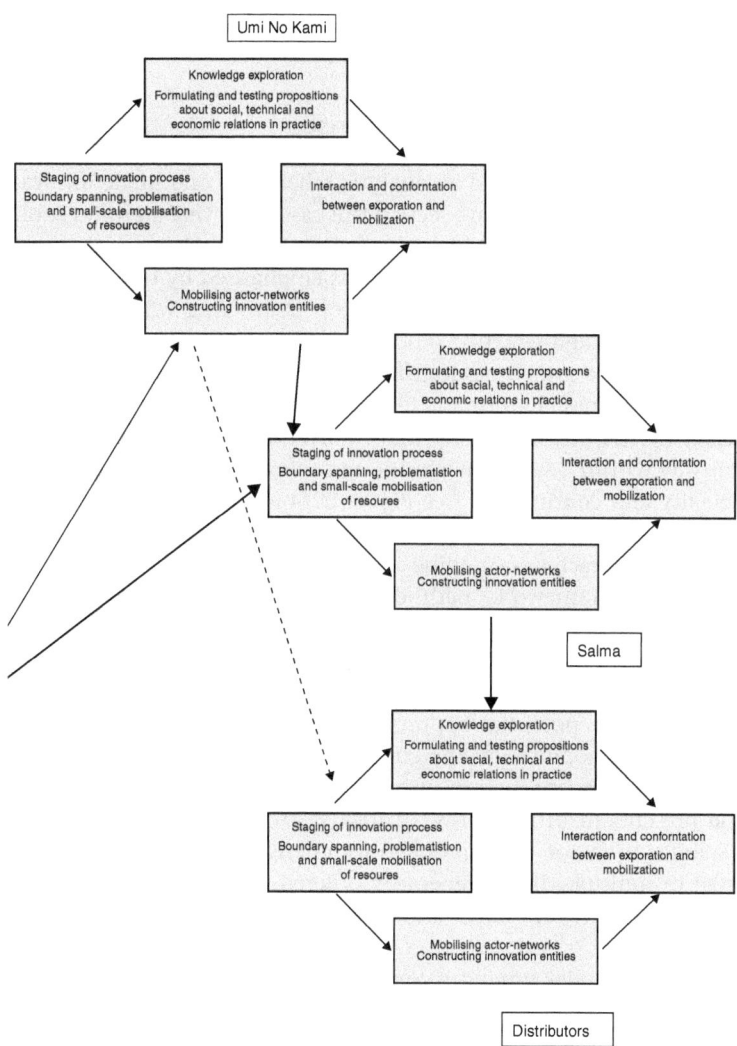

would reveal some of the generative and limiting dynamics of innovation processes; where new meanings are negotiated, choices have to be made and new courses of action pointed out.

A network of interconnected processes

So far, I have sought to build an analytic scheme for one single innovation process. However, we have seen in this case study that an innovation process cannot be isolated, which also follows logically from a relational perspective; in particular we saw this was emphasised by the industrial network approach. The innovation process is evolving within a set of interconnected and interacting processes/actor-networks mutually influencing each other; sometimes translating the other into one's own actor-network, sometimes aligning interests and sometimes taking over and/or betraying the other. The fact that innovation processes are situated in this way within a set of interconnected actor-networks, increases the complexity and contingency (see Figure 22) of the innovation process. Furthermore, it suggests that we also need to observe innovation processes as they interact, (re-)combine and battle with other processes.

Within this framework, path dependence becomes the dynamic effect of the historical, geographical, cultural and material orderings of multiple actor-networks. Is the implication of this that history does *not* matter? Is it the *present* ordering of multiple networks/practices that matters? No, potential options are certainly outcomes of history, but also of the mobilisation/imagination of potential futures (Hernes, 2007). It is thus a negotiation of multiple pasts with multiple futures in the present. In other words, history matters as enacted presence. The stabilising factor – and problem for change – is that practices are enacted in multiple, heterogeneously ordered networks and practices, meaning that both voluntarist and determinist views of innovation are impossible.

The process model produces complexity when including the interaction between several actor-networks seeking to realise different programmes/projects, and possibly seeking to translate each other. This is illustrated here by a version of the process described in this study. It is this interconnectedness among multiple networks and practices that produces (the experience of) path dependence, or friction (see Chapter 6).

I will now turn, in Chapter 6, to the discussion of my case study, using this analytical scheme for discussing existing literature and how my study may contribute to the field of innovation studies. After discussing the overarching dynamic of 'contrary forces' of innovation processes, I move on to discussing my case study based on the different aspects of the analytic scheme. First, mobilisation of actor-networks is discussed; second, knowledge exploration is discussed; and, finally, the interaction between the two sub-processes, as well as the confrontation and interaction between the innovation processes and the network of interconnected processes in which they are situated, are discussed in the last part of Chapter 6. Relevant literature is presented, challenged and complemented in each part.

6
The Contrary Forces of Innovation

This case study is an example of a relatively radical innovation process. The analytic framework outlined in the previous chapter positions knowledge – or perhaps the lack of knowledge – as the central problem of innovation processes. The overall innovation process is characterised by a number of uncertainties that, due to their complex and interacting nature, nobody can predict. When referring to the 'nobody-knows problems' in this chapter, I am not arguing for the ignorance of the practitioners who I have studied. On the contrary, the actors in this case study – as in many cases of industrial innovation – are extremely knowledgeable in their fields. Their lack of knowledge pertains to the connection and translation of knowledge and technology between settings. In other words, it is through the act of crossing boundaries and connecting previously unconnected elements that we can say that 'nobody knows' what it takes to succeed. However, by constructing and amplifying the distinction between mobilisation and exploration in a bipolar model, we can explain some of the micro-dynamics of innovation processes from an angle that has not yet been sufficiently described in the literature. First, during (sub-)processes of *mobilisation*, actor-networks are recruited and committed to things with which they are initially unfamiliar: an idea, a prospect or a prototype of something that may or may not become feasible and useable. Yet, to enable mobilisation, a degree of certainty has to be presumed. Second, during (sub-)processes of *knowledge exploration*, the aim is to create knowledge – to explore the object and its potentials – and therefore change is unavoidable. Moreover, this process of generating knowledge tends to multiply (alternatives of) the object, and hence increase rather than decrease the uncertainty/complexity – or development risk – of the project. Finally, the *interaction* between mobilisation and exploration processes on the one hand, and between different actor-networks/organising processes on the other, often leads to controversies and compromises that may set the project off in new directions. This chapter, then, is devoted to a discussion of the theoretical implications of the conceptual framework when used to interpret the case study. I will contrast and complement the outcomes of my study with theories of path creation (Garud and Karnøe, 2001), innovation as punctuated learning process (Van de Ven et al., 1999) and relational/interactive perspectives of users and markets (Pinch and Oudshoorn, 2003; Håkansson and Waluszewski, 2007a; Kjellberg and Helgesson, 2007b).

The chapter outline is based on the analytic scheme of Chapter 5 in the following way: implications of the model as a whole are discussed in the first section. In the next section I start with a review of Callon's (1986) model of translation and Garud and Karnøe's (2001) model of path creation/mindful deviation, and then go on to discuss the implications of the mobilisation aspect in my analytic scheme. Thereafter, I review Van de Ven et al.'s (1999) model of punctuated learning, before I build an argument for knowledge exploration as a divergent process. In the last section I discuss the dynamics of interaction between mobilisation and knowledge exploration – moreover, the challenge of interconnecting/embedding the innovation process in a web of interconnected processes. 'Users' are used to illustrate this both in theory and the case study. Each of the sections is divided in two main subsections, starting out with critical examinations of existing theory, then discussing the theoretical implications of my study.

Knowledge and the lack thereof is, in several ways, assigned a prominent position in this model of innovation processes; from the outset, we can maintain that innovation processes include multiple 'nobodyknows' problems that cause indefinite development risk. This is neither controversial, nor the main finding of this study, but it relates to the one problem that all research on innovation seems to agree upon: uncertainty. The presence of a number of 'nobody-knows' problems frequently produces high development risk in innovation projects, making it impossible to conduct reliable evaluations or have particularly trustworthy knowledge of them. Even behind seemingly simple ideas, a relatively high level of complexity will be revealed when the matter is investigated. The Neptun/Umi No Kami/Salma projects described in this book were not very advanced or comprehensive in their origin; the idea of fermenting fish and making a salami-like product of fish does not look like something that would revolutionise the fish industry. However, as the number of technical and economic 'problems' to be solved steadily increased, the process grew immensely complex.

I believe that the description in Chapter 4 has produced a picture of the contingencies of such a process: the uncertain elements, the things that became nothing, the things that were brought together and then turned into something, and the great development risk.

The ordering of heterogeneous elements into durable patterns and relationships (Law, 1994) seems a sensible way to describe innovation activity. However, it is important to remember that these terms do not necessarily refer to single actors with clear goals. It may equally be a matter of multiple interacting actors with unclear, multiple and conflicting interests and goals. Nevertheless, I would argue that a successful innovation process can be observed by identifying new and relatively stable socio-material patterns. In industrial settings this is made complex by the intertwining of several interconnected patterns: first, of technological ordering of materials and technologies in the laboratory and, later, the reproduction and scaling up of these patterns together with production workers in factories; second, of product ordering of tuning the technology with the preferred recipe and product features, of designing the visual shape and presentation of the product, and of anticipating and adapting the product to how it is going to be used; and, third, user ordering – getting the product included in distribution networks and making the product a part of users' using practices (e.g., of cooking and eating in this case study).

Innovation processes are, in this book, analysed as emerging actor-networks. Authors of various network perspectives speak of hybrids in various ways, including hybrid collectives and hybrid objects, underlining the description of any object, practice or organisation as a heterogeneous network; a result of its relationships. However, in some settings these network conceptions seem too crude (Hanseth et al., 2004); thus, Law and Mol (2002, in Hanseth et al., 2004) talk about how the technical and social are tightly melded together, almost like mixing fluids. Latour similarly speaks of the 'seamless web' (1999: 204) – that is, in relation to the realisation of extremely complex sociotechnical networks, as in Hughes' study of electrical networks (1983, in Latour, 1999a). Too often, 'networks' are described as more or less depictable sets of nodes and links, while often actor-networks perform unities and entities. While an organisation or an industrial network to some extent is separable in terms of humans, departments, firms, technologies, money flows, production and distribution, etc., other actor-networks appear to be either too complex (even diffuse), as in large information systems, or too unitary (compact, blended), as in the salmon salami in this case study. We are talking about a carefully assembled set of elements, which are made to perform a unity, or to work according to a certain programme. Sometimes it may appear uncertain, ambiguous, in multiple versions or not as an assemblage at all. At other times it may succeed in holding together in a certain shape, being presentable (and rejectable) to users. This process of ontological uncertainty, or we could say ontological flexibility, is due to the uncertain nature of an emerging actor-network (or object) in the making. Many different (social and material) preferences, interests and 'programmes of action' are to be merged or dissociated in the battle for the actor-network's identity and shape. With this clarification of concepts, we can move on to discussing the potential theorising from this study.

Realising and stabilising innovations

In this conception, innovation processes consist of two partly distinct and partly interacting sub-processes of mobilisation of actor-networks and exploration of knowledge. There are often tensions between these sub-processes to the extent that it makes sense to talk about them as contrary forces. Whereas mobilisation is directed towards aligning interests and reducing risk, exploration is directed towards formulating and testing propositions about reality. And whereas mobilisation seeks to converge, exploration frequently leads to divergence of the innovation. Thus, the interaction between these could well be described as controversy, or battles over meaning and resources. Those responsible for generating knowledge – making the innovation work both technically and economically – will often argue for adding more resources, including more time to explore potential solutions, more expertise, more equipment and new raw materials. Those responsible for mobilising the actor-network and the resources, on the other hand, will be more concerned with making the idea communicable (i.e., simple and clear), as well as reducing risk through minimising investments, ensuring 'progress', stripping down much of the complexity and adapting to existing practices and networks. Mobilisation requires boundaries, as it will always be easier to mobilise resources related to specific and concrete goals. This is not to say that mobilisation and exploration are always handled by different actors, but rather to clarify how these are based on opposing principles and logics, which thereby often conflict and challenge each other. Sometimes it appears almost like the innovation process is at war with itself, as the internal tensions between these kinds of sub-processes tear and rip at each other, with one tearing the other apart. Still, occasionally the bits and pieces may start to connect to each other and bring the innovation significant steps towards real-isation. The framework in Chapter 5 should also be viewed as being circular, in the sense that, in most cases, it will require several such processes, each having different goals and programmes, before achieving something like a 'successful' innovation. Only in very 'simple' cases or in cases with improbable luck can we think of an innovation process having only *one* mobilisation and *one* exploration. In addition, to add to the mess, this contrapuntal process is always situated within a network of interconnected processes, in which some things are made possible while others are not.

Putting knowledge, and the lack thereof, at the centre of attention, the framework suggests that innovators have to produce two different kinds of knowledge: first, a chain of arguments suited for convincing, mobilising, maintaining and removing (parts of) actor-networks and their resources; second, testable propositions about reality – for example, regarding how to make the technology work and what users have interest in such a product. Innovation processes are propositional at their core. The original idea is a proposition about the potential that stems from a new combination of elements. This idea needs some resources to get started, and then the idea needs to be explored in practice – testing whether and how the proposition may hold. This will normally happen by breaking the original idea into a series of new and 'smaller' proposi-tions; as the innovation is opened up and investigated, it is revealed as a more or less complex set of problems, all having many different solu-tions *in potentia*. However, to enrol allies, it is necessary to make the idea and concept converge on a number of aspects, and this will often cre-ate a 'lock-in' for the for the remaining process. To an extent, the con-verged version of the concept binds the exploration part of the project to what has been prospected, and one will often have to let things run for a while before it becomes possible to change things. An example is when Tine's top management locked the project team of Umi No Kami for an extended period of time within a certain frame; a blended white and red fish recipe, international before domestic marketing, etc. This could not be changed before reconfiguring the set of actors, resources and ideas involved, both within management and in the team. In addi-tion to the obvious political reason of aligning various interests with a certain conceptualisation of the innovation, another reason for this 'lock-in' is that it is difficult to make choices about a new thing without knowing 'enough' about what it is. Thus, there is need for exploring and testing the present version before changing it. When the direction of the case study project was finally changed, the inclusion of pre-rigor raw materials was used both to provide new faith in the sausage and, at the same time, to expand the 'horizon of opportunities' as the involved actors saw it could be possible to do other things with this new resource. A third reason is the lack of capacity to cope with the level of complexity involved in innovation processes; hence, decision-makers will need to cut things down. In this sense, it seems very hard to take a 'rational management approach' to innovation, as the uncertainty is too overwhelming. By cutting things down and making choices in the face of uncertainty, things frequently go wrong. On the other hand, the actors would perhaps not do much better in increasing complexity either.

Some of the problems characterising innovation processes are highly uncertain; they are 'nobody-knows' problems'. Thus, the innovation has to be taken through a number of phases, where some problems may be solved simultaneously while other problems require a sequential order, and every such 'nobody-knows' problem' has its own set of phases. In other words, I would argue that every 'nobody-knows' problem' of an innovation has to be taken through several stages to be resolved, and that the complexity – and hence risk – of innovation processes is equal to the number of 'nobody-knows' problems times the number of stages they have to be taken through. In addition, we have to add that these problem-solving processes are connected to each other in various ways; they are intertwined and interdependent, and changes of one aspect influence other aspects. The resources to be connected are themselves complex, and in addition have to be connected in complex patterns with a number of other resources within the process. Moreover, the innovation has to be adapted to the wider network in which the process is situated. The explorative process is expanding, as it meets new connections, creates new opportunities or meets friction from established practices, always in the space between creative imagination and empirical testing. Hence, we see how unwieldy innovation processes may become. Such innovation processes produce a multitude of options. Many disappear, others are put on hold, and a few are realised.

When new elements (human or non-human) are enrolled and mobilised in the innovation process, they often seem to have more influence than those already in place in the process. This happens sometimes to confirm and stabilise aspects of the process, as when Nordvi was approached by Slinde and proteins were included to stabilise the fatty acids, when the 'right' bacteria culture' was included to make the fermentation process work optimally, or when Swensen was hired and 'near-infrared' scanning technology was included to control fat content in the recipe. During and after the implementation of these elements, they were described as crucial for making the innovation work, and their successful integration in the process dominated the description of the innovation process – until a new element had to be included and hence took over. At other times the effect was more dramatic in terms of altering or disrupting the process, as when Skjervold, Swensen and their colleagues were hired by Tine R&D, and their introduction of pre-rigor salmon from Bremnes Seashore ended up taking over, translating, the entire fish salami project, and made it into a fresh loin project instead. From this, I would argue that there is a relationship between resource mobilisation and time: elements enrolled and mobilised in the innovation process tend to have more influence on the innovation the later in the process they are included. The reason for this effect seems twofold. First, the element representing the most recent changes – improvements or alterations – has more narrative power in representations of the innovation process, such as project reports and presentations, in their symbolic power as representatives of 'progress' in processes of high uncertainty and unclear evaluation criteria. Second, in order to mobilise new actors and resources into the process, sometimes the significance of the results already achieved has to be downplayed in order to increase the necessity of the new. Argumentative power may be produced by increasing the importance of the new relative to the old. To establish understanding of the benefits of using fresh pre-rigor salmon, the effects of the previously successful testing and choice of bacteria cultures were downplayed, as technical process, resulting taste, texture and durability were brought in as central parts of the argument for including an exclusive and expensive raw material and its owner – a small family-owned fish farm.

Another take on this would be to view exploration as a process of expanding and generalising the concept. Here, the concept has to be brought towards stabilisation, and it must be possible to generalise. This presupposes that the concept is tested in practice; that it shows it holds. Such concept generalisations in industrial innovation often involve hypotheses about appropriation and scale. Appropriation refers to the ability to mobilise, control and internalise resources and relationships. Scale refers to the idea that elements of the concept, such as technologies, products and users can be generalised – that is, high volumes of the same product can be produced, standardising technology and translating the products into massive use (by a few 'large' users, or many 'small' ones). If, as is often the case, the concept does not hold in some aspect, it has to be adjusted to achieve the supposed effect, then recombined with other elements, and then again generalised and tested in practice.

The need for cost absorption and therefore generalisation relates both to concept and technology. This happens certainly in standardising technology to scale up the production volumes and thereby cut costs per item, but also in expanding the brand with more product variants, using the developed technology in several settings. As mentioned previously, the exploration process is not a one-way street, as the innovation and its elements often 'speak back' at the innovator (Latour, 1996). Whether talking of 'material agency' (Pickering, 1995), or of social and material resistances (Law, 1992; 1994), it is clear that not anything goes. The question of success in knowledge generation activities is not only a question of 'objective' experiments by skilled experts, nor simply a question of the political struggle over meaning and vision. This case study, and numerous other ethnographic and historical studies of science, technology and industry, show how this is also a matter of heterogeneous interaction. The objects under investigation and development tend to 'speak back', showing different interests in and capacities for recombination than anticipated by the innovators. It was hard to keep mould away from the production facilities of fish salami; eradicating it required considerable adaptation of the production, during a six-month 'negotiation' process in the laboratory, and then a somewhat shorter period in the production facilities at Bremnes. Fisheries supplying white fish would not immediately adjust to the requirements of Tine R&D, and the salmon supplier (Bremnes Seashore) also had other (economic) interests at play when sorting, prioritising and processing raw materials for different customers. On the market side, they had to interact closely with consumers in the supermarkets, demonstrating the product, to get them to even consider buying the salmon salami. Moreover, the proposition of the 'dream customer' – restaurant and catering actors – was unexpectedly and brutally abandoned by the MRC corporation in Asia. This exploration process, of knowledge generation through proposing, testing, interacting and adapting (aspects of) the innovation, is supposed to move asymptotically towards solid knowledge. Further, as can be seen between the lines in most of this text, knowledge is here nothing more – or less – than stabilised relations, between texts, technologies, raw materials and social actors of various kinds. Here, the paradox is that while the aim of exploration processes is to produce knowledge, they (almost) always produce complexity; multiplicity, ambiguity, choice.

If the knowledge sought through exploration processes is defined as (temporary) stabilised relations, we can immediately see that the problem of complexity – the complexity produced by the exploration process – is of a relational matter. This is where the mobilisation process comes back in. Eliminating or reducing development risk means reducing the number of associations and relations involved, and simplifying the innovation and its actor-network. This happens sometimes by black-boxing the actor-network involved, at other times by cutting off connections and associations to force the process in a certain direction. This strategy of 'radical simplification' becomes particularly visible – and crucial –when the innovation process moves closer to commercialisation, looking for and adapting to users. The idea that one should 'start with the market' does not hold when creating something new. As the case study showed, it is not easy to involve just anybody in such a process, the innovation is not stable enough, and will not have power to resist pressure from sceptical actors; only very incremental innovation would be possible. If the concept is unstable and contains many variable elements, it is likely to generate a high number of new concept alternatives, which are difficult to handle in practice. Hence, a period of exploration is necessary, to see whether and how the elements connect, even though this will in itself produce more complexity before the mobilising forces get back in to provide simplification and choice. From this, I argue that, for the stabilisation of the innovation and its actor-network in cases of radical innovation, one needs to achieve a certain degree of stability of the innovation before involving users. If not, the concept will not be able to resist pressure from sceptical actors. Moreover, with a high number of unstable and variable elements in the concept, interaction with users will produce a number of new possibilities that are impossible to handle in practice. Therefore, when finally going for something, one tends to go for the alternative with the fewest (new) elements, the most similar alternative to what exists from before. When going to users and markets, the question is more a matter of reducing risk, as well as eliminating and stabilising elements and relations, than it was during earlier stages of the process, and few new elements are easier to handle. All of the 'nobody-knows' problems involved, and all the elements to be connected, cannot be isolated; they are not independent. Hence, it is difficult enough with the simplest version of the innovation. So, when commercialising a 'final' version of the innovation, one goes with the alternative containing fewer elements, which is most similar to the existing, to reduce development risk. Commercialisation of radical innovations, therefore, consists of a process of radical simplification.

In addition to interacting with each other, these processes of mobilisation and exploration interact with what has long been known as 'path dependence'. In the economic and historical literature, path dependence has been used to describe how historical contingencies, often accidental events, may create more or less irreversible 'lock-ins' and 'dominant designs' that limit and determine the subsequent direction of the industry (Garud and Karnøe, 2001). However, within the framework suggested in this book, path dependence simply means that every actor-network is situated within a historical, geographical and cultural setting within which everything is not possible. Actors are limited by their previous relations, experience, investments, etc. Path dependence may be related to local and global political regimes, which are different between agro-food and seafood, different technical systems, different market networks, different cultures, all of which then influence production practices, cooking and eating practices, etc. For the most part, path dependence has been described in the literature as being a barrier to innovation, 'framing' innovation processes and forcing companies to act rather conservatively within, or close to, the actors' existing set of relations and practices. For example, Bremnes Seashore was not able to economise on their technical innovations within their existing marketing practices, their existing distribution network and due to their location within a 'spot-price' market for fish as raw material. However, this also illustrates some of the structural bias within path dependency literature. Dependencies are largely seen as structural and therefore rather static. I argue that such dependencies are not at all static and given, but dynamic and emergent. In a process view of innovation, we need a dynamic view of path dependence that explains how dependencies come about, change and disappear. The 'market system for fish' was not static. It is true that it changed slowly, and it was not possible for Bremnes Seashore (at least not in the short term) to change the way fish was marketed and distributed in that system. Yet it was certainly changing, albeit gradually, and the successful domestication of salmon has led to large consolidations of fish farming companies. Increased control of the raw material supply and quality may in turn lead to ability and interest among the companies to change their practice to be more in line with the agricultural industry, towards more industrial activities like product development and branding. From a process perspective, path dependence is better seen as a relatively slower process, maintained via carefully and often long-term assembled and intertwined networks of heterogeneous elements; creating what Håkansson and Waluszewski (2001a; 2001b) have called 'friction'. The change in one resource combination is also likely to lead to change or resistance in other resource constellations. 'Dependence' thereby becomes a relative measure of one actor-network's stability (in time/durability and space/distribution) in relation to a number of other connected actor-networks or processes. Hence, instead of serving as an explanation of how innovation is hindered by some rigid structure, this view explains some of the slowness and some of the unexpected outcomes of innovation processes: (1) why innovation processes tend to take significantly more time than expected, and (2) why 'successful' innovations often are realised as incremental changes or additions to the existing set of relations. Path dependence in this version may also explain why and how some kinds of innovations are possible within particular settings – for example, it was possible for Tine to invent and commercialise new food products within their setting, due to being a central part of a 'heavy' techno-economic system able to handle technical development, distribution and marketing of differentiated food products. However, when crossing sectoral boundaries and venturing new business between agri- and aquaculture, it was no longer obvious how to innovate, and whether they would succeed or not. Path dependence, in this version, includes all developments that take place around the actual project, and sometimes such 'friction', as noted by Håkansson and Waluszewski (2001a; 2001b), may even be an important enabling factor for the innovation process. In some fortunate situations, there are specific developments and movements in the wider network of the innovation process' that are compatible with the innovation, and which therefore may help the innovation towards its realisation and use. On the research side, the development of 'blue-green' innovation politics was important for the mobilisation of funding from the Research Council, and the emergence of a biomarine innovation strategy within the Tine corporation was central in the decision to buy the patent application from Professor Slinde. On the market side, there were fewer such coinciding and compatible movements, both in the fish industry and in the agro-food industry, although it is not unlikely that such a movement towards industrialisation, product development and branding of fresh fish products is on its way.

Mobilising and committing to uncertain outcomes

'Mobilisation' is located at one of the poles in the framework. Power to mobilise resources and decisions for innovation is produced by constructing a (more or less coherent) chain of arguments. Ironically, this coherence has to perform as a degree of certainty; the mobilising actors need to assume that they know what cannot be known. In line with the concept of 'translation' (Callon, 1986; Latour, 1988), this means that parts of the innovation process are based on a relational logic of 'power production' – that is, carefully building (or connecting to) networks of human and non-human elements with interests in realising the innovation. It is about developing conceptual and material actor-networks that are able to translate to other actors, including securing their interest and mobilising them. Plans are written, such as project applications, business plans, market strategies and progress reports. Rituals are activated, such as estimations of 'total markets', which involve taking the idea through formal bureaucracies of middle managers and committees, or advocating the project in front of the board of directors. All of this has only one purpose: to mobilise actors and resources. By building a trustworthy and coherent chain, and developing the elements within it, the innovation may gain the power to convince others. However, the mobilising actors are rarely allowed to build their empires of meaning without disturbances. An unexpected outcome of the exploration process, a disloyal ally, or the requirements of a new partner, may cause the actor-network to shake or collapse and stimulate the intense activity of enrolling and aligning actors to save the project. Hence, mobilisation processes happen in several rounds, never reaching a final and stable state. I will here discuss how this aspect of the mobilisation/exploration framework contributes to the study of innovation, complementing previous studies within STS (Callon, 1986) and innovation management (Garud and Karnøe, 2001).

Translation and path creation

According to Callon (1986), the art of building actor-networks, of gathering bits and pieces into a unity, is the work of translation, or the work of how actors create alignments. The scheme developed in his article for telling these stories of alignments divides processes of translation into four 'moments': problematisation, interessement, enrolment and mobilisation (ibid.). *Problematisation* is the work of both formulating questions to be investigated, and determining and defining the identity of the social and material actors to be involved. The aim is to shape an 'obligatory point of passage', through which every involved actor must pass to formulate its identity and role. This is done by convincingly demonstrating that all actors' interests are served by admitting and joining the proposed project. A problematisation describes a system of alliances in the face of a problem, thereby defining what these actors 'want'. After making this 'crack' in the present reality, the next step in Callon's model is *interessement*, or to impose and stabilise the identities of the other actors, who can either submit to the proposal, or refuse – for example, by instead accepting the identities offered by 'competing' networks. In this sense, the existence and influence of other networks and relationships that may influence the emerging actor-network in question is part of the model, but I would argue that there is a need to expand on this aspect to cover the complexity of an industrial setting. While problematisation is often formulated in and via texts, such as project proposals, interessement involves testing whether these proposed identities and relationships hold in practice. An infinite number of techniques and strategies may be deployed to achieve this, such as bringing in various types of evidence, in addition to negotiation, force, seduction, etc. The aim is to realise a system of alliances, as described in the problematisation. The outcome of interessement is called *enrolment*, or the successful distribution of roles, producing alliances of a more stable kind, via 'multilateral negotiations' and 'trials of strength'. However, it is not easy to distinguish enrolment from interessement in the model, and it could probably be viewed as being one and the same. The final step in the model is called *mobilisation*, which means 'to render entities mobile which were not so beforehand' (ibid.: 216). The crucial question in order to realise the research programme is: who speaks in the name of whom? Who represents whom? A few involved actors are made to represent many, from generalisation and assumed representation, but will the masses follow their representatives? As has been described in studies of many difference scientific disciplines, this is mainly done through series of translations. In Callon's study of domestication of scallops, the scallops were translated into larvae, and then into numbers, and then tables and curves, thus becoming easily transportable to project meetings and scientific conferences. Enrolment is thus transformed into active support. This support is then communicated through 'spokesmen' – network members reduced to representatives for its network. However, Callon warns us, enrolment is rarely easy, as every process and every actor can break down or betray the enroller, and the ability to mobilise resources and forces is crucial (e.g., money, threat or just 'good' arguments). Actor-networks are fragile and temporally limited constructions that may be contested. 'Closure', a state of relative stability, occurs when the spokesmen are deemed to be beyond question, taken for granted, and the controversies along the way forgotten, deleted or 'black-boxed'.

In the case of Umi No Kami/Salma, the 'problematisation' happened gradually, through many rounds, and was a growing process involving an increasing number of actors, thereby leading to a reformulation of the proposed problem several times. It appears that industrial innovation processes are more complex than typical scientific processes, having to go through many episodes of problematisation to take the innovation through its many stages before perhaps achieving commercial use in the end. This involves both continuity, including gradually adjusting the formulation of the problem and the goals of the project, and discontinuity, as sometimes dramatic changes may occur in the 'fundamental' aims and rationale for the project.

When comparing this with my observations, Callon's scheme seems to imply a coherence between programme and practice that is not always there. The problematisation (the programme/text/idea) and the 'realisation' in practice (what is tested and constructed) may be more or less unrelated, and may not interact for long periods at a time. Hence, it seems that the interaction and confrontation between them is typically triggered by certain kinds of events (e.g., when resources, such as time or funding, start running out). Callon's model may also be accused of allocating too much power to the innovators. It is certainly true that sometimes competing actors and networks are dissociated, as is the case in some instances in my case study. Still, in industrial settings, innovators usually seem to be forced to adapt to others, adjust their program towards the existing set of associations, rather than dissociating and outmanoeuvring them (Håkansson and Waluszewski, 2001a; 2007a). Callon does not, in this article, develop any account of how the actors are forced to reformulate and transform their own programme in their interaction with others.

In an industrial setting, the stabilising issues of representation and closure need to be challenged on two points. First, representation is rarely located in one place. Representation of – or ownership of – the meaning of the project seems in general to be more temporally limited, more contestable and more distributed in industrial settings than is implied in Callon's model. Second, 'taken-for-grantedness' is not enough in industrial settings to bring about 'closure', as the final word in industry is not said by a community of professionals (such as researchers, technologists, managers or even marketers), but by (commercial) users, actively acknowledging the innovation's value (such as in a new product) and using it in their daily practice (Pinch, 2001; Håkansson and Waluszewski, 2007a).

As I have tried to show, the differences between science and business call for elaborating on – or changing – Callon's model, to enable study of industrial innovation processes. When scientists approach their ideas, a whole apparatus of norms, methods, funding bodies and networks of colleagues are orchestrated along the lines of a well-established rationality, including writing research proposals, recruiting allies and doing research. Ideally, this is done in a rather linear fashion, although most of the time even science emerges in non-linear ways, resulting in a need to alter things along the way, as has been solidly documented by STS scholars. On the other hand, when business actors seek to realise innovations, research is just one of several parts of the process; the main parts of the process involve many other actors with other norms, practices and rationalities, such as managers, marketing personnel, technologists and designers, distributors, various customers and, in the end, consumers. Moreover, while 'success' in science may be to produce 'negative' knowledge – that is, discovering that something is not possible – 'success' in business may only be claimed if (1) the innovation actually works, and (2) it is being used in a way that is seen to create value of some sort. I would claim that this makes industrial innovation processes more complex (e.g., through the number of involved actors and elements), more uncertain/contingent and more pragmatically related to method than in science. Here, rigidity may be more strongly related to commercial goals. The connection between 'problematisation' (posing and mobilising actors for an idea) and 'interessement' (actual exploration and testing) often becomes looser, but also potentially more dynamic, as more alternative futures are possible (due to pragmatism). In addition, stronger confrontations between interests may occur (increasing risk), and more radical reformulations of the problem/aim may be enforced.

A somewhat different way of conceptualising innovation processes, particularly related to the mobilisation aspect, has been provided by Garud and Karnøe (2001). This conceptualisation partly builds on later ANT/STS research on technological and economic practices, such as Callon's argument for 'framing'[1] of markets (1999), and connects this to innovation studies within management and organisation. They oppose the concept of 'path dependence' in industries (e.g., David, 1985), which explains the emergence of novelty from historically contingent processes and often accidental events. This may create more or less irreversible 'lock-ins' and 'dominant designs' which limit the potential for action, and hence determine the subsequent development direction of the industry. Path dependence is accused of granting too much determinacy to history, and hence giving little or no room for agency of entrepreneurs participating in shaping future history. In this perspective, the emergence of novelty is serendipitous (Garud and Karnøe, 2001: 7). Instead, Garud and Karnøe argue for what they call 'path creation', or investigating the role of and space for agency in shaping new industrial practice, or, in other words, of shaping new paths. The concept of 'mindful deviation' is developed in an argument for how (collectives of) entrepreneurs work over time to realise and implement new ideas in the economy. In sum, entrepreneurship is human agents' 'ability to span boundaries of relevance structures, translate objects and mobilise time as a resource' (ibid.: 25), and in this way 'disembed from existing structures defining relevance', and 'mobilise a collective despite resistance' (ibid.: 2). This is not a unidirectional movement, as 'actors interact with one another to negotiate the relevance of objects and behaviours that constitute the technological field' (ibid.: 10). Such mindful deviation opens up options, not only in the initiation of the process, but also in the continuous 'ability to create and exercise options' (ibid.: 7) as a crucial part of generating momentum, switching to new alternatives along the way that may have greater promise than the original idea. Path creation often depends on the co-evolving of several groups, in processes that both may constrain and enable each other.

Garud and Karnøe identify cognitive limits to the path-creation process, as actors are deeply embedded in technical fields and economic limits, and sometimes they need to exploit what has already been created (2001: 11). However, less is said about limitations to knowledge, such as the feasibility of future technical, economic and social relations. The innovation process will succeed, only if the 'right' actors and resources are mobilised and (social) resistances overcome. However, this is contingent, first, because different actors and resources may be 'right', meaning that they generate momentum and progress, albeit in different directions, and, second, because materials (technologies, raw materials, texts, etc.) also have 'agencies' (i.e., characteristics or preferences) and may thereby generate resistances to the project. Consequently, in their case study of 3M Post-it Notes, Garud and Karnøe refer to human resistance, while overlooking the potential technical (molecular or other) resistance towards the effort to enable particular use and/or persuade particular users in favour of the invention. Still, the exploration of materials and economies, and the overcoming of material and economic resistances, also require imagination, manipulation, negotiation and skilful combination of social, material and economic elements.

The argument for viewing path dependence and path creation as 'two sides of the same coin' (Pinch, 2001; Mouritsen and Dechow, 2001) is consistent with a process view, in the sense that history always matters, both as a constraint and enabler for path creation. As Garud and Karnøe (2001) and others have argued, we need to study how social, cognitive, technical and institutional matters co-evolve and intertwine

in the process of creating new paths. Consequently, one should endeav-
our to study the interaction of different actors and actor-networks
(with different interests) *in the making*. What, then, are the dynamics
of industrial innovation processes and stabilising innovations? How are
they 'achieved' when trying to innovate (stabilise new patterns) in the
face of uncertainty? Although it is not implicated in the model, Garud
and Karnøe still seem to emphasise continuity over controversy, hardly
demonstrating how competing interests may provoke confrontation and
how the mobilisation of time or the translation of objects may be contested issues. In their aim to conceptualise agency, the model also has a
social bias, in comparison with its partial roots in actor-network theory,
as if everything is possible regarding technical matters, and that it is
only a matter of social agents' imagination and mobilisation. Moreover,
the ability to mobilise 'molecules and minds' (ibid.: 18) does not mean
that these elements will turn out to be combinable in the long-run. It
simply means that a space for exploration is made, but there is still a
long way to go before succeeding in creating and stabilising a new path.
Hence, I argue that the mobilisation/exploration framework might help
to better get a grip on the interactions and confrontations between the
mobilisation and exploration process, including testing whether and
how social and technical relations may be possible to realise in practice.
This is what I will now turn to.

Mobilisation in the face of exploration

In expanding on 'mobilisation' with clear parallels to both Callon (1986)
and Garud and Karnøe (2001), my study shows how the presence of
unlimited uncertainty, and the interaction with 'reality' (i.e., the exploration process of formulating and testing propositions of reality) have
effects on the mobilisation process. Busch (2004; 2007) has recently
argued that distribution technologies and systems (supply chain management) – and the actors in control of these (a few big globalising retail
actors) – are taking gaining power in the food industry. This underlines
the importance of including the mobilisation aspect, and the politics
of the industry, in order to understand commercialisation dynamics
within this industry. Here, it becomes crucial to ask certain questions,
such as: who sets the rules of the trade, who gets access to the marketplaces and what effects does it have? My study sheds light on this aspect
by describing attempts at entering various marketplaces with a 'frame
breaking' product concept. Actor-networks such as distribution systems
may or may not have influence on particular practices and projects.
Distribution and market systems within the fish industry, the agro-food

industry and even within the pharmaceutical industry have had direct influence on the outcome of Tine's blue-green projects (both enabling and resisting). At times, they have also had a role in 'staging' such activities, contributing to change and stabilisation of frames (Callon, 1999; Garud and Karnøe, 2001) and evaluation principles (Garud and Rappa, 1994; Stark, 2006), thus also influencing where opportunities are sought out (Beunza and Stark, 2004), and how it would be possible to exploit them – whether influencing technoscience (blue-green innovation), labour costs (taxation and trade regimes), access to marketplaces (supply chains/retail companies/restaurants) or other.

The expert-actor, Slinde, was in position to 'stage interaction' between some previously unrelated elements from aqua- and agriculture, as he had expertise within both. This idea of combination was arguably not random – it was deeply based within Slinde's repertoires and networks of knowledge as an expert. It also did not have to be these specific elements, as there would have been a number of other possible combinations of elements. We can rather say that the idea was contingent, both based on the repertoire of the actors coming up with the idea, and on their preferences, assessments and access to resources. It was based on a specific repertoire of knowledge, enabling a certain degree of creative recombination of elements, *a priori* assessment of these ideas, and access to facilities where they could be tested out in practice. Such a recombination of elements from different fields of practice is often explained as coming from *boundary-spanning* individuals and groups (Orlikowski, 2002), or via *boundary objects*, including artefacts of various kinds, such as information systems, theoretical models, prototypes, etc. (Star and Griesmer, 1989). While, in this case study, Professor Slinde most certainly acted as a boundary spanner, he did not span random boundaries. He spanned boundaries between his own fields of expertise, moving between familiar areas, yet nevertheless areas that previously did not interact much. His particular interpretation of the message of 'blue-green innovation' was about translating and transforming fish into a resource for industrial production and marketing, similar to that of meat and poultry. This version of the blue-green programme was not necessarily what the emerging blue-green actor-network of institutional actors had in mind. As we saw in the early reports from the Research Council on their blue-green research programmes and projects, Slinde's interpretation was barely represented in this actor-network at all, while several other versions had more momentum among the more central participants. However, in the staging-process, Slinde still had to rely on – while at the same time subverting – existing frames of reference: fermenting meat/fish the same way as when producing salami. We could say that he started the recombination process based on two simple hypotheses: (H1) fish may be fermented the same way as meat into salami and (H2) such a hybrid product should be able to find use/users. Such industrialisation had been possible with other materials eaten by humans, and both meat from mammals and poultry has been industrialised on a large scale with great success. Now, with the break-through of aquaculture, the turn towards industrialisation had come to fish. Slinde associated with the blue-green programme/actor-network, but also deviated from it, via his radical interpretation – seeking to transform fish into an industrial resource in line with agricultural prac-tices. It is a process of 'mindful deviation' (Garud and Karnøe, 2001) to explore and shape new frames of reference, and create new industrial paths; a creative process, not starting from nothing, but using elements from one actor-network to challenge and translate another.

The initial material experiments with the idea could not, however, prove certainty – neither of technical feasibility nor of commercial potential. The inherent ambiguity/uncertainty of combining the raw material of fish with technology for curing of meat was reinforced, lead-ing to discursive uncertainty in terms of the interpretation of the object, what it could be used for and by whom, and material uncertainty in terms of how to stabilise the object technically. Still, after 'dressing up the bride', presenting the miserable first versions of the object for repre-sentatives of business (ForInnova) and science (the Research Council), it achieved enough interest and economic funding to let the project con-tinue with new experiments. The 'fish salami' had passed its first trial.

Rather weak signals from various sources had to be 'read' together to create an overall picture of this being worth pursuing. Actors with resources (funding, knowledge, facilities and ingredients) had to be interested and mobilised on behalf of the idea. Still, how and if the technology would become feasible was uncertain. How a nutritional product could be shaped out of the technology was uncertain. Its poten-tial use and the interest of users were as uncertain. Finally, it was funda-mentally uncertain which actors would be interested in industrialising and commercialising something from this, while also having the ability to do so. By adding the unpredictability of interaction, or of how oth-ers respond, studying innovation processes from a relational point of view expands the sources of uncertainty, compared with seeing oppor-tunity, knowledge and markets as objective entities waiting for discov-ery. Van de Ven et al. (1999: 170) make a similar argument, focusing on systems or community-level factors as a main source of uncertainty, hence proposing that 'more novel innovations require greater change in all system functions and, therefore, greater development time and greater chance of failure' (ibid.: 171). However, interactions during an innovation process are likely to happen between all kinds of actors and resources, including humans, organisations, materials, technologies and other.

When Tine R&D began its cooperation with Slinde, the aim of building up competence on protein applications was a goal that was present in the project most of the time, not just in the project application to the Research Council. In practice it was a genuine interest of the involved project participants and in discourse perhaps also as a back-up in case the commercialisation of fish, or the fish salami to be more specific, did not work out. What also seems to be a common attribute of entrepreneurs is the ability to jump between arguments according to what connections to make with whom, who to convince and what resources to mobilise. Van de Ven et al. (1999) describe how aims and strategies, both in top management and in the operational running of innovation projects, changed over time, according to the phase, status and needs of the project.

In the process of selling the invention to Tine, we can see a set of discursive strategies used to make the idea interesting and adaptable to the buyer's setting. Since the idea had not yet achieved any degree of stability, Slinde and his associates had to work on how to formulate it. First, it was presented not as a consumer-ready product, but rather as an *opportunity*. The buyer would secure the rights to exploit a potentially useful technology for fermenting fish. Second, the presentation of the invention shifted between emphasising what was specific and what was (potentially) general. The invention was presented as specific in referring to 'fish salami', and how such a product could compete either for the consumers who do not normally eat much fish, or for conscious fish eaters wanting a less greasy alternative to smoked salmon, or even for highly demanding and innovative Asian consumers. The invention was presented as general when emphasising that this patent would give the owner the rights to potentially an entire portfolio of cured fish products, of which the salami was just the beginning. By stabilising a specific representation of the object as a fish salami, it could be presented to potential buyers as something concrete, which could be developed and commercialised within a reasonable time-frame. At the same time, the object was kept open for interpretation,[2] making adaptation to the buyer's setting easier, and also presenting the object as an opportunity much bigger than just a strange combination of fish and salami. Third, the context of the idea was framed in a particular way. The (Norwegian) fish industry was framed as totally incapable of industrialisation, lacking competence in R&D, distribution and marketing, and being embedded in a market system that does not reward innovation. The agricultural food industry, on the other hand, was framed as very capable, potentially being a serious challenger of the fish industry related to product development and finding new ways of distributing and marketing fresh fish (hence creating economic and use value). This rendered only a small handful of large food producers able to exploit this opportunity.

As described above, Tine, after some consideration, decided to buy the patent application from Slinde and ForInnova, and immediately started the product development project Umi No Kami to develop and commercialise a fish salami. When it happened that the Tine management wanted to go for this opportunity, and to a large extent buy into the 'frame' offered by Slinde and ForInnova, it was not only because of the sales effort of ForInnova, with help from Slinde and Nordvi; rather the fit with Tine's emerging biomarine strategy, coming in so early in its development, was crucial. At this point in time, Tine was willing to take some risks, being conscious about the need to learn about the fish industry, and still was not experiencing strong commercial pressures. Moreover, this project took part in shaping Tine's biomarine strategy by involving many people and resources in the organisation.

The Umi No Kami and Salma projects were not without controversies; few projects are. There have been conflicts between researchers and business management, between inventor and buyer, and between developer and producer. One way to 'read' the processes that may illustrate the relation between mobilisation and politics on the one hand and knowledge exploration and generation on the other is via the concept of 'evaluation principles' (Garud and Rappa, 1994; Stark, 2006). Different actors evaluate – assign value to the object – by means of different criteria, or principles. When exploring the innovation, researchers might value an object (or technology) in terms of quality standards, functionality and problem solving. Marketing people might evaluate in terms of (anticipated) customer needs and preferences. Business managers might evaluate the innovation in terms of revenue and margins, economic criteria and fit with the overall strategy. And the supplier/producer (Bremnes Seashore) probably evaluated the fish salami in relation to the prices for their raw materials, people employed in their local community and opportunities for entering new market systems. These different evaluation principles are not (always) complementary, as they are often rooted in – or co-producing – various actors' interests. Sometimes they overlap, sometimes they are in conflict and often they emphasise completely different qualities and characteristics of the project. Yet this plurality of evaluation principles, which creates ambiguity with regard to the innovation, is necessary for innovation to happen. At the same time, it demands processes of mobilisation, including negotiating, deleting and combining evaluation criteria and interests. In to this perspective, politics are a way of naming an aspect of what inevitably happens in every innovation project. Some views are bridged and combined, some are 'winning' battles and others are left out. While other ANT-influenced perspectives on mobilisation (see above) have explained this negotiation with existing (and often competing) actor-networks via concepts of 'mindful deviation', 'problematisation' and 'interessement', there is another dynamic that also deserves some attention during mobilisation processes: the *exploitation of uncertainty* in interaction between (potential) allies in efforts to stage and mobilise resources for innovation.

The existence of knowledge 'fields' and 'practices', as well as 'frames of reference', means that different actors have different experiences and expertise related to the characteristics (and potential) of the elements recombined into a new idea. Moreover, different actors are situated within different sets of relationships – for example, with (potential) users, thus having different knowledge of what it takes to mobilise a network strong enough to take the innovation through technical and economic trials on its way to feasibility and usability. However, few – if any – know what it takes to relate previously unrelated elements to each other. Simply, nobody knows, but some know more than others, and this may be used by innovators in the process of mobilising actors and resources. I argue that there are three aspects of the relationship between lack of knowledge and the mobilisation of actors and resources for realising innovative ideas (the building of actor-networks) that need to be explicated. First, asymmetrical knowledge and experience may be used during interaction to exploit the other. In the process of building arguments to convince others on behalf of an innovation, presumptive competent actors are mobilised to represent the innovation as something worth pursuing, whether they possess technical expertise, or are market actors, users or something else. Second, manipulation based on asymmetrical knowledge consists of mobilising apparent authority. Hence, if more 'radical' ideas will be impossible to evaluate in objective terms, those with more (or different) experience with (some of) the elements involved might be able to exploit actors with less or other types of experience. This is the case with all parties involved, both with actors buying the idea, investing partners and customers of various kinds. Third, asymmetrical knowledge and mobilised authority may therefore contribute to the construction of a 'market for ideas' by those who have 'knowledge advantages' – that is, a trigger for economic exchange of ideas. Choices regarding innovative ideas always have to be made while in possession of too little knowledge.

In order to illustrate this argument, let me recall two significant events from the case study. First, Professor Slinde had long-standing experience with research from both agriculture and aquaculture. He also had experience with patenting and commercialising new technology within aquaculture. Based on his unique combination of academic and business knowledge from different sectors, he came up with the idea of fermenting fish in the first place. This experience was also embedded in relationships to various actors, like research institutes, funding bodies and industrial firms. In these relationships, he could relatively easily mobilise some 'expert' authority to rally economic support and technical cooperation in order to start the process of exploring whether and how this idea could be feasible, first in a pre-study, and then in a full-blown research project. In the next round, Slinde understood that it would be too difficult for him to be in charge of the industrialisation and commercialisation processes, so he instead sought to sell the idea to an industrial actor. He did not have the industrial organisation and resources that were needed to scale up, market and distribute such a product. He also did not know anything about the challenges of industrial competence in the fish industry, related to establishing a high-quality supply of raw materials and high-quality production routines. However, his professorial knowledge and authority did not suffice in selling the idea to a commercial actor. The technology transfer office, ForInnova, was enrolled for this purpose, and took responsibility for co-representing the idea with Slinde. Furthermore, the idea was dressed in the formal authority of the stamp from the patent office. A recipe/description of the technology became associated with the patent institution, which was registered for the public/official quality control of new ideas. However, this did not mean that the idea became stable, as they could not know if there would be objections from other inventors, or if the patent office would approve the application as innovative enough, etc. In addition, it is likely that the less one knows about patenting, the more it becomes possible to overestimate the stability of this element. Tine, which was already involved with R&D people and resources, was interested but hesitant regarding buying the patent application. At this point in time, Tine had modest knowledge about the fish industry, fish as raw material and about patenting; they had just started their involvement in a few projects related to biomarine resources, and were involved in a learning process about such matters. Further, due to their dominant position within a protected national market, they had done very little to build competence on patenting. This increased their feeling of uncertainty, in addition to the inherent uncertainty regarding both the technology and use of the innovation. It was only when another commercial actor was mobilised by Slinde/ForInnova to represent the idea as commercially interesting (the Japanese actor that was to consider buying the patent application), that Tine's management decide to go for it.

Second, when the increasing impatience and the reorganising of Tine's biomarine portfolio enforced and enabled a reorientation of the Umi No Kami project, new knowledge and authority were mobilised to gather support for a new direction. The recently hired group of researchers from the University of Life Sciences brought with them expertise and established relations to another innovation; pre-rigor processing of salmon. The combination of these ideas – about fish salami and pre-rigor salmon – produced new enthusiasm for the Umi No Kami project in Tine Ingredients; it represented new opportunities both for technology and for expanding the product concept. This new constellation of the business unit management (Hovland), R&D people (Skjervold, Swensen, etc.) and the fish farm (Bremnes Seashore) together built a renewed programme for the fish salami project: an exclusive brand based on a pure pre-rigor salmon product. In addition, a large French catering actor was recruited to represent the project from the market side. With new expertise, new resources and new relations, Hovland succeeded in mobilising Tine's owners to once again provide their full support through approving and investing in the new programme. While the uncertainty of the technology and the product had been reduced, the uncertainty regarding users and markets was still huge. Soon after the Tine board of directors approved the product, the French actor was off the hook, but they had already served a part of their function in contributing to a convincing argument for the new direction of the project. The recruitment of Bremnes Seashore into this 'new programme' was also clearly based on differences (asymmetries) in knowledge. In being quite desperate for finding ways to get returns on their R&D investments, the people at Bremnes had great hopes for Tine's vision of selling huge amounts of salmon salami worldwide. They showed great faith in Tine's ability to transform their commercial knowledge/expertise from agro-foods to fish products, not realising that most of Tine's commercial practices were deeply embedded in relationships with other actors and networks that might not necessarily be transformed to serve a new kind/category of products. Hence, the question was whether Tine could translate Bremnes Seashore and their supreme processing technology and raw material into their 'fish salami' actor-network. On the other hand, as the last part of the story showed, the pre-rigor salmon ended up translating the Tine project into serving Bremnes' agenda, betraying the salami while maintaining its actor-network, transforming the project into a fresh salmon loin project.

Furthermore, the 'coup' of the project by means of a new constellation of raw materials, researchers and arguments, not only meant that resistance within the project had been eliminated; in the same move, potential resistance from top management had been avoided too, backed up by a 'promising' relationship with a prominent French actor. Armed with these new arguments, or we could say this new *vision*, the director of Tine Ingredients had a seemingly easy task of convincing Tine's top management of the new direction of the project. Yet, this is also an important competence – knowing how to convince, how to enrol those in power, secure control on the main direction of the project, ensuring better access to necessary resources in the future. In such situations, it seems out of the question to reveal doubts and 'too much' of the apparent uncertainty of the proposal. This competence of the director was mentioned by several of the project participants. This may be insignificant to my argument in itself, but when contrasted with top management's lack of approval of suggestions proposed by the previous project group, particularly with regard to the white/red fish mix and technological investments, it shows how important every link in this complex emerging web may be for the outcome of the project. The search for industrial partners had been going on, both on the fish and the meat side, for a long time, but as soon as new evaluation principles, or 'frames of reference', had been approved, all alternatives were deleted in favour of the supplier of the pre-rigor salmon – a small fish farm. The argument was that they needed just this kind of raw material, and that production facilities should be built as close as possible to the raw material (i.e., by sea), ensuring freshness and quality of the fish. Hence, this became more important than collaboration with actors that had competence in industrial food production and/or an established distribution network for similar products.

What are the consequences of this? What can be done in the face of unlimited development risk? It seems that the kind of uncertainty involved in such processes enforces learning. Most of the lacking knowledge cannot be purchased; it is not available, since no one has done this before. The necessary knowledge has to be generated through active learning, and to enable this one has to mobilise actors, resources and time in the face of uncertainty. This is one answer as to why innovation is difficult, and the more radical, the worse the situation is. In retrospect we can see that Tine bought something – an idea and a patent application – that turned into nothing, at least within the time-frame of this study. The idea of fermenting fish/making fish salami never stabilised conceptually. The technology was taken through many trials and finally stabilised, but it did not find commercial users. The meeting with unlimited development risk created far more problems than was (or could be) foreseen, both technically and commercially.

Constructing an actor-network around a new idea is an exercise in creative connecting of actors, networks, resources and ideas. In this sense, this part of the process does not concern itself with 'truth'; rather, it is about producing power effects – that is, mobilising actors and resources on behalf of the innovation – and translating their interests into a common project. Still, if and when an actor-network is mobilised, the elements employed in the chain of arguments may produce frames and evaluation principles that define the room for action in the project, hence enforcing temporal lock-ins that cannot easily be broken out of in the subsequent parts of the innovation process. Therefore, the way in which this process is handled is not indifferent. Elements from the exploration process, such as hypotheses and propositions about the outcome products and markets, and relationships to partners with valuable resources, are 'borrowed' in the building of a coherent chain of arguments and a powerful actor-network. On the other hand, the case study also showed that after a while it became possible (and necessary) to renegotiate the framework conditions, and that the actors could resell the project with brand new conditions to the top management, by bringing in new elements and actors, partly from the proposition/knowledge generating/exploration process and partly from loose connections to potential users.

The mobilisation and the exploration (sub-) processes are different in the sense that they only partly connect and interact. Sometimes they operate on their own and according to their own logics, but then one, for various reasons, may be confronted by the other – for example, due to impatience from the mobilised actor-network regarding the (lack of) progress in the exploration process, or because there is a need to add resources to the exploration process, which forces the actors into new rounds of mobilisation. Thus, the sub-processes draw on, but also challenge each other during their execution.

Exploring knowledge and generating complexity

'Exploration' is the process of actually testing and developing the innovation. Such exploration processes are typically about imagining and formulating testable solutions (e.g., in the form of theories or propositions for problems of a technical, social and/or economic character) and then testing these solutions in practice. Viable, or stable, propositions are thus the outcome of analytical conceptions and empirical interactions. Loose ends have to be tied into a concept that is testable in practice, in an interactive process of tuning the concept with its social, technical and economic relations. Is it possible to combine fish and fermentation technologies? Can a mix of white and red fish produce a high-quality product? Can whey proteins stabilise biomarine fatty acids? Will Asian and European customers be interested in using the product in their cooking and eating practices? How should it look, taste and smell? Would industrial users and/or consumers use it, and would they be willing to buy it as a 'high-end' product? Can the production be scaled up at Lofotprodukt, at an abandoned meat facility or at Bremnes Seashore? The list goes on. This is a kind of process that is similar to the scientific method, that is, the interaction between intellectual imagination (Weick, 1989) to formulate 'propositions' about reality, and then test them out. Loose ends have to be tied up, creating concepts that are testable on technical combinations, calculations and users. It may also be likened to engineering methods of carefully assembling various elements into entities and relationships that work, of making the innovation technically and economically feasible.

Exploration and mobilisation are partly separate sub-processes, and the case study shows how these processes often followed completely different logics and, further, that they often did not even deal with the same object. Examples of this are when the mandate was to make a mixed white/red fish product, while parts of the project organisation succeeded in proposing and testing a pure salmon version. Another is when research funding was mobilised for investigating and designing milk proteins for various applications, while the Neptun project largely worked in preparation for producing a very specific product: a fish salami. Further, there is the instance when the commercialisation manager gradually moved from marketing the fish salami – the product he was hired to commercialise – to marketing the raw material instead. The innovation (and its (potential) relations to use and users), as it was described during the activities of mobilising resources, decisions and allies, was often different from how its realisation was sought during technical and market development activities. Much of the time, the R&D projects and, later, the marketing activities went on exploring fermentation and protein research, market research and conceptualisation, etc. At the same time, the project's representatives in Tine's management went on doing their mobilisation work to maintain goodwill and commitment to the project, by presenting reports, emphasising progress and downplaying or explaining reasons for delays, and giving a general impression of economic potential. I will here go further into the dynamics of exploration/knowledge generation and its interaction with mobilisation, suggesting that my study complements Van de Ven et al.'s (1999) cyclical model of innovation.

Punctuated learning

'The Minnesota Innovation Research Project' (MIRP) (Van de Ven et al., 1999) conducted a seminal study that had a great impact on our understanding of innovation processes. Analysing their comprehensive data materials Van de Ven and colleagues drew on Anderson and Tushman's (1990) 'punctuated equilibrium model of cyclical change' that described how industry-changing technological breakthroughs were followed by convergent and incremental movements towards a 'dominant design' (Hargraves and Van de Ven, 2006). From this, Van de Ven et al. (1999) could recognise and conceptualise a pattern of the innovation process as 'a nonlinear cycle of divergent and convergent behaviours that may repeat itself over time and reflect itself at different organisational levels' (Van de Ven et al., 1999: 213). They found this to be the case independent of the great diversity in paths and outcomes in the processes studied. Linear stage models, as well as random models, are opposed; instead, they argue for innovation as 'emergent process' based on nonlinear dynamics, in which sensitivity to initial conditions and the ability to manage complexity (metaphorically described as navigating in the river rather than controlling it) were viewed as being crucial to success. Hence, innovation is viewed as a learning process, with 'learning by discovery' being 'an expanding and diverging process', and 'learning by testing' being 'a narrowing and converging process' (ibid.: 203). Thus, these two processes are viewed as being dependent on each other in a continuous cycle. However, less is said about the mechanisms, or micro-dynamics, that initiate and fuel the different processes of divergence and convergence.

Similar to the observations of my study, Van de Ven et al. (1999) underscores the fundamental uncertainty of innovation processes, as 'the usefulness of an idea can only be determined after the innovation process is completed and implemented', and such summative evaluations 'are not available to the managers and entrepreneurs who are undertaking the innovation journey' (ibid.: 11). In addition, they found that the evaluation criteria – the principles for assigning value to the innovation – used by innovation managers and resource controllers shifted over time (ibid.: 42), in line with the changing needs of the innovation process and the unexpected events that occurred, related to outcomes, process and input. Such changes 'triggered innovation managers and entrepreneurs to search and redefine their innovation ideas and strategies' (ibid.: 42). Beunza and Stark's (2004) concept of 'evaluation principles' in identifying innovative opportunities from boundary crossing activities and Howard-Grenville and Carlile's (2006) concept of institutionalised rules in cross-domain innovation have provided similar insights, but from a more political point of view. They all show how the negotiation of evaluation criteria is fundamentally a political process through which power relations are (re-) constituted, and that 'borrowing' evaluation criteria across boundaries is one of the keys to identifying and realising novel and valuable opportunities.

A central problem from this intersection of the fundamental uncertainty of innovation processes on the one hand and shifting evaluation criteria on the other is what Van de Ven et al. (1999: 12) call the 'management of paradox', which means that highly effective organisations are able to perform 'in contradictory ways to satisfy contradictory expectations' and 'ambiguity in goals'. One such paradox, which is incorporated in my process framework, is found between mobilisation and exploration, hence offering insights into some of the microdynamics of how and why divergence and convergence come about during innovation processes. When discussing the handling of – the managing of – innovation processes, Van de Ven et al. (1999: 124) argue that, in uncertain situations, there is a need for 'a pluralistic power structure of leadership', increasing the chances for technological foresight, while, however, also decreasing the chance of oversight. The diversity in views and conflicts is seen as constructive in ensuring good process during divergence, serving as 'checks and balances with each other', while unitary, single-vision and hierarchical leadership tends to restrict creativity and deviant behaviour. From this, innovation processes are described as steadily producing new 'spin-offs' – new ideas and projects – taking the process in new and often multiple directions, where it is difficult to predict what will succeed. Nevertheless, as I have argued above (in Chapter 6, section 1), convergence, in the shape of 'radical simplification' (e.g., through vision or force), is necessary to mobilise (and sometimes to maintain) actor-networks and resources. Van de Ven et al. (1999: 50) further describe how, over time, 'more and more players are brought into the game', resulting in a complex network, 'engaging in a series of transactions necessary to move the innovation forwards'. However, I would argue that including more players is likely to bring about divergence instead of clear and converging solutions, hence enforcing political processes of mobilisation when 'progress' is needed. Sometimes this might be done through aligning the interests of multiple participants, while at other times it is necessary to cut off relations to 'troublesome' actors and resources, hence reducing rather than expanding the actor-network.

Lastly, Van de Ven et al. (1999) found that 'innovation uncertainty decreases over time as system functions that define key technical and institutional parameters for the innovation emerge' (ibid.: 172). However, in my case study, the reduction of uncertainty was not only about getting system functions or institutional arrangements in place, nor was it about technical and social relations stabilising 'by themselves'. Rather, it was about processes of mobilisation, negotiating interests, involving strategic partners and especially radical simplification – of stripping down the innovation itself in order to get adaptability, and thus momentum – towards relations and patterns of distribution and use. Sometimes the innovation and its actor-network were simplified to an extent that, in reality, transformed the innovation. This is a key problematisation of the outcome of the Minnesota Innovation Research Project: what are the conditions for, and the dynamics of, reducing uncertainty in innovation processes? Moreover, how are the divergent process fuelled to the extent that convergent processes are so desperately needed to take the process further?

My study of innovation processes between agri- and aquaculture was not designed to test or compare with Van de Ven et al.'s (1999) model. Nevertheless, it is an in-depth study, more ethnographically oriented, and with similar points of departure – to develop an empirically grounded understanding of innovation processes. In some sense, my study confirms and aligns well with their *Innovation Journey*, but, perhaps more importantly, I identify, elaborate on and provide rich descriptions of some of these mechanisms through which innovation processes are organised.

Exploration as divergent process

What seems to be less covered in Van de Ven et al.'s (1999) model, as well as within the dichotomy of path dependence and path creation, is the process of exploring and constructing new knowledge – of formulating, building and testing knowledge in practice. When actors and resources have been mobilised into a new innovation project, a process of *making things work*, or *testing propositions in practice*, starts. While these aspects have been studied more or less as one and the same within actor-network theory, I have observed in my fieldwork how these kinds of processes do not always interact. Sometimes they are not coupled at all, and when they do interact, tensions, confrontations and conflicts are often produced. They tend to challenge each other, because it is extremely hard, in innovation processes, to make the ideas and meaning produced when mobilising actor-networks and the actual results of development activities come together. As mentioned previously, these seem to be two relatively distinct kinds of processes – or activities – and therefore deserve more attention than they received in the path-creation model, and they are not as tightly coupled in practice as is suggested in Callon's model (in the assumed coherence between problematisation and interessement/enrolment). In practice, this tends to produce several parallel paths, as also demonstrated by Van de Ven et al. (1999), which sometimes interact and confront each other. I suggest that the reason for this is that the exploration process, of generating knowledge, challenges the mobilisation process that is already established, and creates a dynamic that either may strengthen the emerging path, or undermine and destroy it, or even lead to the establishment of new path-creation processes entirely. While they are certainly more or less intertwined in practice, these processes are distinguishable analytically: mobilisation is a process of getting and maintaining the rights, alliances and resources to go on with the innovation process, while exploration is a knowledge-generating process of proposing and testing out relations in practice.

As described in the research application to the Research Council, the Neptun project had a research design common to many technology development projects, a process of modelling. The building of a 'quantitative model' (model 1) – that is, a theoretical representation of the invention – enabled scientific testing of a physical model (model 2), a prototype, and having the results of various versions of the prototype analysed in scientific terms, providing results that again could be fed back to – and adjust – the quantitative model. This was an interactive and mutually constituting process of developing a technology and scientific knowledge about that technology. The technical problems the project sought to resolve were about stabilisation of micro-biology: getting a firmer texture without fat slipping out of the product, and improving durability (slowing down the process of oxidation of biological materials/fat). In addition, on the side of the main research problems, issues related to fermentation (testing bacteria cultures), colour, hygiene and raw-material quality had considerable focus from the project team. It is interesting to note here how the aesthetic ideals for the modelling of a prototype (model 2) had strong associations to salmon, while the use of salmon at the same time increased one of the main technical challenges in the project: stabilisation of fatty acids. The resulting colour of the model recipe was somewhere between grey and red, according to the amount of saithe (or other white fish species) and salmon, but the ideal was seen to be a clear salmon red colour. Various ways of achieving this, from adding artificial colours to increasing the content of salmon, were tested. When the project failed to finalise the stabilisation of a fish salami (although taking several steps in the 'right' direction), it was to a large extent seen to be due to problems with the white fish; of getting fish with good enough micro-biological quality – or in other words of getting fishermen and fisheries to handle the fish with the necessary hygienic standards.

Similar practices of natural sciences have been described by several STS researchers (Latour, 1999b; Knorr Cetina, 1999), showing how relations between the theoretical modelling of scientific knowledge, and practical interaction with 'phenomena in the world' (nature, technology, etc.) are made possible through systematic 'chains of translation' (Latour, 1988). However, what seems different in an industrial setting like this, is how this 'chain' goes a few steps further; the result of such technoscientific exercises has to be taken far beyond scientific journals, through industrial production and out to supermarket shelves and consumers' situations of use (kitchens and food plates). In order to be deemed as valuable, the Neptun and Umi No Kami projects would have to prove useful in enhancing or creating economic value somewhere in the value chain. Therefore, this temporal stabilisation of the innovation, which was useful for scientific procedures, could later turn against the project, ultimately hindering or even betraying it, due to the rigidity that such (mental and physical) reifications of the emerging object might imply. One out of many possible versions was chosen at a certain stage of the process, without anyone knowing whether this version had the properties needed to survive later trials of different kinds – conceptualisation, market research, customer adaptations and use and so forth.

Knowledge is situated, and hence difficult to transfer across settings (Orlikowski, 2002). Yet, it is still relevant *who* is trying to move and translate knowledge. It seems that some actors are more likely to succeed than others, depending on the – sometimes unexplored – compatibilities between the specific knowledge regimes represented. In this case study, we observed how the actors started exploring and developing new cross-domain practices in buying a patent application and then seeking to develop the invention based purely on their own practice. New resources and new technologies were brought into an existing knowledge regime, and development was pursued within an existing system of interconnected practices. Evaluation principles (Beunza and Stark, 2004), or conventions (Howard-Grenville and Carlile, 2006), from their existing socio-material practice of processing milk, related both to microbiology and to categories of users, were applied to find feasible and valuable solutions. Tine took for granted that they could get what they lacked in knowledge by interacting with others. However, they had not foreseen that they could not utilise or connect with the other field of practice without first learning more about that practice themselves. Ironically, the act of protecting knowledge by organising the project in-house had the unintended consequence of losing important knowledge embedded in the relations of academic technoscience. The technical problems that arose after moving the experiments from the Food Research Institute to Tine illustrate the embodied nature of knowing in practice. On the other hand, we cannot be certain that the presence of someone used to working with meat would have helped the translating of the practice to fish. Anyway, this independent strategy of innovation did not work out when working on materials from another industry. For their own knowledge to become translatable, they needed more knowledge about the practice of 'the others', and therefore they started recruiting allies on the fish side.

Actor-network theory does not say much on a principled basis about what the (material) practices of actually doing expert work consist of. It is wisely left as an empirical question; the surgeon performing a skilled procedure, the biologist getting bacteria cultures aligned in a fermentation recipe, the production worker keeping nasty micro-organisms away via hygiene routines and even the marketer, surveying consumers, interacting with – and selling to – customers. In such empirical accounts from actor-network theory authors there are many descriptions of professional practices, but there seems to be a lack of distinction between attaining the right, the position or the resources to undertake a practice, on the one hand, and actually undertaking a practice on the other. To me, these seem to be markedly different kinds of activities, even if they sometimes interact during the overall process. In this particular case study of an innovation process, attaining the right and the resources is what I call 'mobilisation', and the *undertaking of the practice* I call 'exploration'; that is, knowledge generated by ordering elements in durable patterns. Mobilisation activities may, for example, be expressed via problematisations, suggestions and promises like: 'let's check it out', 'this is a problem (or an opportunity), and I have the ability to solve it', 'let's do this together, and you will be rewarded', 'give some more time, money and resources, and I will show you that it will work' and 'there must be users, somewhere, for this product'. It seems for the most part to be a convergent activity, in the sense that a number of discursive techniques, like argumentation, vision, simplification, promise and visualisation, are employed to achieve support for initiating and maintaining the process, until it is realised something that may become durable 'on its own' (i.e., by maintaining its own relations, or by being maintained by others – users of the innovation).

Exploration activities, on the other hand, may be expressed more via propositions and statements about potential sociotechnical relations, that in turn are tested, redesigned and tested again, in attempts at ordering social and technical elements into relatively durable patterns. Examples of such propositions from this case study may be the following (in my wording): 'it is beneficial both from an economic and a technical point of view to mix white and red fish in the salami recipe'; 'if we make a salami of salmon, then these particular kinds of users will buy it'; 'if proteins are added to the recipe, then the fat will stabilise better'; 'if we improve the hygiene routines, then the mould will go away'; 'if we can identify the best bacteria tribe, the product will stabilise and taste better'; and 'if sell fresh loins under the same Salma brand, it will be embraced by high-end customers'. However, exploration is not only about making and testing propositions, as the object (and other involved elements) frequently 'speaks back', interacts with the other involved participants, and may thus cause change of direction, resistance to certain ways of going about it or triggering new ideas. This is why exploration processes, or knowledge generation, somewhat counter-intuitively tend to be divergent, and produce a multiplicity of opportunities and potential futures. In their passionate engagement with objects, experts tend to find a number of interesting problems, opportunities and potentials. This is, according to Knorr Cetina (2001), a fundamental characteristic of 'epistemic objects' – that is, objects under investigation by experts. They are objects 'characterised by a lack of completeness of being', and therefore tend to expand and multiply.

The learning outcomes of the exploration processes, as well as the confrontations between mobilisation, exploration and other actor-networks, often create disruptions and divergence. This illustrates and elaborates upon Van de Ven et al.'s (1999) description of 'punctuated' learning during innovation processes. Sometimes positive or negative discoveries from the exploration of technical and economic aspects of the innovation lead to smaller adjustments and improvements to technology and concept, while at other times the implications of the knowledge produced fundamentally undermine or alter the project. More often than not, the exploration process seems to produce choice, including choices between solutions, conceptualisations, users, partners, etc. Sometimes learning happens incrementally, as small adjustments to the concept, a simple reassurance to the management or minor increases in economic and time-frames, typically ensuring continued effort to realise the innovation. At other times, learning means departing from the original idea, creating a mismatch to the extent that a battle for the future direction of the project becomes inevitable. What may have seemed to be a stable and consensual actor-network thus dissolves into smaller constellations with different interests, in a battle of strength for achieving the right to further define and participate in shaping the project. In order to understand the development of new knowledge and new practices, we need to relate the process to its frame of reference. When such frames change, what counts as knowledge changes too. In other words, we talk not only about change *within* the existing discourse, but also changing the discourse itself. This, however, will not always happen. Sometimes things are developed or exploited that are already part of present practice, and change in such cases would take place on the premises of the existing discourse or practice, and hence change reality (knowledge/practice) to a very modest degree.

When every new solution brings with it a set of new challenges, it is not easy to simplify and converge. One example is the efforts to minimise the problem of the high and variable fat content in salmon, where a mix of salmon and white fish was proposed to do the job. However, white fish brought with it problems of micro-biological quality and fishermen and fisheries' practices and routines, in addition to unsatisfying colour. Solving the colour issue by adding colouring substances would bring challenges of nutrition and regulation. Solving the quality issue of supply proved hard in the face of an old historically and industrially embedded practice. In addition, a number of different white fish species were tested to find the one best suited in combination with salmon. Another example is the continuous interaction and paradoxification between technical development and commercial conceptualisation. The choice of raw materials for the recipe was not only a matter of technical feasibility, but also a question of anticipated users. On the one hand, a mass product would need low-cost raw materials and could possibly do with a lower experienced quality of the product, which matched the original idea of a mixed product. On the other hand, a product for high-end segments would require absolute top quality in the user experience of colour, texture, taste, packaging and concept, which seemed to indicate a pure salmon product. However, the exclusive use of salmon in the recipe was regarded as impossible technically; in addition to considerably increasing the costs, it was outside the framework conditions in their mandate from the Tine management to make such an alternative at that point in time.

In the previous section on mobilisation, I discussed the political aspects of uncertainty. Yet the lack of knowledge is more directly a problem calling for exploration, as active learning processes to generate knowledge. The exploration process is driven towards divergence, and thereby produces more rather than less complexity. To sum up the exploration process, we see that the number of technical and economic issues that need solving increase as experts dig into the innovation (epistemic) object, and, therefore, the number of possible/potential concepts, products and users are kept high. We also saw in the case study that, after a successful mobilisation, the resulting framework is kept tight until new confrontations and reconfigurations enforce or enable renegotiation either of the framework conditions or of the actual innovation. When trying to cope with the increasing complexity of exploration processes, we saw that participants in the Umi No Kami project repeatedly avoided involving experts from other organisations, such as the Norwegian Food Research Institute and the University of Life Sciences, despite needing their competence (as in the problem with mould in the Tine R&D laboratory). We also saw how, after the 'coup', the new constellation dramatically downsized the project organisation in the process of shaping the Salma concept, despite losing knowledge both on the technology and of potential markets. Finally, we saw that Tine avoided involving industrial users (i.e., distribution actors) for long periods of the innovation process. On the other hand, if they had chosen to include more knowledgeable people and more resources in the development, this would likely have influenced the process, perhaps in ways that were incompatible with the interests and ideas of the participants. Such interaction avoidance may have a variety of reasons, such as fear of loosing control and strategic positions in the project, unwillingness to compromise on some professional principle or lack of knowledge on how to facilitate the interaction process.

In any case, interaction avoidance seems to be a common challenge of innovation processes, basically stemming from the need to handle and reduce the divergent and expanding aspect of exploration. I would therefore argue that innovating actors tend to avoid interacting with others during exploration processes because of the risk of being influenced – or even taken over by – the others. Furthermore, when interaction happens, it may lead to more or less radical shifts in the innovation process, like the multiple 'spin-outs' in Van de Ven et al.'s (1999) model. New connections, new elements interacting with and confronting the project create new conceptions and solutions, that in turn require new processes of mobilisation and exploration. Thus, I suggest that interaction during exploration tends to produce divergence, sometimes leading to shifts in the process; new connections, new confrontations and new elements may trigger new directions in, and spin-offs from the emerging path. To make it even worse, as mentioned in the first section of this chapter, the value of elements that later enter into the process (which turn out to be combinable) seems to have more influence than elements connected earlier. Hence, later relations may become decisive for what happens next and for getting support. More is at stake, and therefore more of the previous elements have to be downplayed in order to strengthen the mobilisation power on behalf of the new elements to be recruited. The original actor-network may thereby become victim of others' interests. At the same time, such interactions and adding of elements may be crucial for the fate of the project.

This interaction dilemma is positioned in the tension between control and involvement. From the project participants' point of view, it is not always easy to make judgements about the balance between the risk of being influenced by others and the benefits of making use of their knowledge and resources. They do not know beforehand whether the project – including their own interests – will be strengthened, or whether they will end up being enrolled and translated into the others' actor-network. Clearly, this fear of being influenced – or even betrayed – by partners was one of the reasons for isolating the Umi No Kami project when the patent application was bought by Tine, cutting off relations to partners that had been contributing to the Neptun project, and that could have helped out – for example, with the mould problem that occurred soon after. There were also numerous references to turf battles and conflicts of interests internally in Tine and in the project group that related to what knowledge to generate, how and based on whose expertise. In the later evaluation of potential partners for scaling up production, they clearly preferred partners that were interested in sharing ownership of the project rather than those only willing (or able) to do the job on an outsourcing basis. They wanted partners that were committed to the project as defined by Tine. Throughout the innovation process, from Neptun, via Umi No Kami, to Salma, there was an escalating battle between the representatives of the different ingredients in the recipe regarding explanations of – and solutions to – various technical issues. This climaxed in the battle between pre-rigor salmon and fish salami, with both sides seeking to translate the other into its project; one of commercialising a fermented fish product and the other commercialising pre-rigor salmon. While this battle was, to an extent, fought on 'mobilisation arenas', there was also a crucial competition regarding the formulation, testing and adjusting of propositions about technical and economic relations in practice. The inclusion of pre-rigor salmon in the project turned out to represent a major shift in the project, with consequences exceeding the calculations of the involved actors. Through a process of testing pre-rigor salmon in the recipe, and adding supporting technologies (such as near-infrared scanning), explanations of the fermentation process were generated that strengthened the power of the new constellation relative to the original project team. Still, the pre-rigor salmon could not take over the project completely before the fish salami had been tested and refused by a set of retail, restaurant and consumer users.

Even if sometimes the knowledge-generating process succeeds in providing clear and singular answers, as in the testing of bacteria cultures for fermenting the fish salami and the use of whey proteins to stabilise fatty acids from fish, most of the time the object and its complex of (potential) relations in this case study expanded during exploration. In fact, the 'lack of progress' in the Umi No Kami project, which was explained by Hovland and Kiland as mismanagement and poor communication, probably had more to do with being in a phase dominated by exploration activities. Hence, they had the dilemma of when to intervene from a mobilisation point of view: when has enough knowledge been generated and stabilised to make choices about how to commercialise the innovation? This was even more complicated by the innovation being unstable within several dimensions at the same time. Hence, the innovation consisted of many sub-processes (both of exploration and mobilisation) that needed stabilisation on their own terms while at the same time having to be combinable with the other elements and sub-processes. New solutions and strategies may appear in many places, sometimes gradually undermining the whole process, while at other times suddenly connecting and binding things together. In this particular innovation process, *commercial* users of various kinds were not involved before this major shift took place, contrary to the mantra of 'customer-driven innovation', which was also popular in Tine at the time. For this reason, the involvement of commercial users – the testing of the concept on potential customers – may serve as a main issue when discussing the innovation process and its interaction with other actor-networks/processes in its environment in the next section.

Interacting and mutual translation of interests

The processes of mobilisation and exploration, as well as the interaction between them, are making this complex enough. Still the innovation process is not evolving in isolation. It is situated in a network of more or less intertwined processes, each with different interests, aims and problems to solve. In this network of processes, the innovation either has to be adapted to the established patterns and practices of production, distribution and consumption, or the network has to be adapted to the innovation.[3] Research on industrial networks (Håkansson and Waluszewski, 2001a) shows how the former is clearly preferred; however, some kind of network reconfiguration and redefinition seems necessary for anything new to be included. It is this interaction between actor-networks to which I will now turn. Van de Ven et al. (1999: 148) reported that development patterns of inter-organisational relationships followed 'a multiple, parallel progression of numerous bargaining, commitment, and execution events throughout the temporal duration' of the relationship. Moreover, interaction in networks was often more influenced 'by activities occurring in other dyads than by the internal logic of the dyad itself' (ibid.: 148). While not limiting the discussion to the interaction between (formal) organisations, I maintain that stabilisation of innovation processes largely happens by embedding them into heterogeneous networks across organisational and other boundaries.

A common limitation of organisational ethnographies, also within the STS literature, is that they receive a too narrow focus. It is hard to observe things happening outside the local interaction, or rather, the interaction that takes place in a very limited set of locales.[4] This is also, to some extent, a limitation of this study, as it is a study of a particular set of related projects. Things happened elsewhere that were difficult to observe from my viewpoint, and still I argue that one cannot understand this particular set of processes without also taking a broader look at the situation. Hence, I have tried to include at least some of the actors that were in direct interaction with these projects, such as the management in Tine and in Bremnes Seashore, the Norwegian Research Council, the government and a set of potential customers.

In the framework (Chapter 5), this is conceptualised as a 'process situated in a network of intertwined processes', thereby providing an angle on path dependence as being emergent and relational. If, in following the framework, we then assume that the innovation process under investigation evolves in interaction with other processes seeking stabilisation (and expansion) through mobilisation and exploration, what can we learn about the interaction between the innovation process and its environment? First, it adds to the already massive complexity and uncertainty of such processes. Not only are the elements of the process to be mobilised and adapted to each other, the innovation process also has to mobilise and adapt to a number of other processes going on. This strengthens the need for partial stabilisations and simplifications (as discussed in the first section of this chapter) before and during interactions with present and (in particular) new relations. Second, it bears some implications for the handling of established and new relationships in industrial networks. I will here discuss the dynamics between innovation processes and the network in which they are situated, suggesting that this may add to our understanding of the realisation and commercialisation of innovations. In particular, conceptions of 'users' and 'markets' for innovations are problematised, and it is argued that there are more kinds of 'users' involved than normally included in the literature. Furthermore, I pinpoint some problems both with reconfiguring established relations and with recruiting new ones in economic/industrial settings. Below I present some recent research on how users and markets are (re-) configured, before I outline some theoretical implications of my study.

Interacting with users

In STS studies there has been a growing concern for the *user* of technology, realising that it is difficult to understand technology development fully without including the receiving end. The findings of these studies have been that users are not only receivers; they are also participants – in more or less direct ways – in shaping the technology. The recent emphasis on use and users has led to insightful studies of technology, both in relation to economy, industry, social practice, individual users, politics and the technology itself (Pinch and Oudshoorn, 2003; 2008). Von Hippel (1998; 2005) has shown how some innovations come about from users identifying their own needs and trying to develop their own solutions. In developing the influential term 'lead-users', he has subsequently researched and experimented with ways of systematically including users in the innovation process - for example, via 'tool kits', workshops and joint research projects. However, Hoogma and Schot (2001, in Oudshoorn and Pinch, 2008) remind us that not all users are innovative, and hence that one needs to think through how to foster a 'sensitive interactive environment for the adaptation of some radical new technologies' (ibid.: 543). They also problematised the concept of users, as many (end-) users remain silent under the self-interested representation of companies, institutions and individuals (e.g., children being represented by parents, government institutions and pharmaceutical companies in medical questions). In addition, I suggest that more kinds of users than just end-users should be included in the analysis of innovation processes.

The social construction of technology approaches have positioned users as *participants* in technology development, in terms of how they interpret and therefore use the technology. The boundaries between users and designers, production and consumption, are thus blurred. Closure of the 'interpretative flexibility' of objects can be reached through several 'closure mechanisms', stabilising the technology in a pattern of predominant meaning and use (ibid.: 544). Designers, users and intermediaries do, however, interact within a 'technological frame', providing institutionalised rules on how to interact with the technology. Users (social groups) and technologies are seen to be co-constructed, but the literature so far 'has not paid as much attention to the diversity of users, the exclusion of users, and the politics of non-use or restricted use' (Oudshoorn and Pinch, 2008: 544). Feminist studies of technology have further emphasised the diversity of users, and encouraged investigations of the relative power relations between designers and various user groups. In this there is an explicit critique of the actor-network approach for putting too much focus on producers and experts, and hence failing to understand non-standard positions.

Within actor-network theory, the concepts of 'configuring the user' and 'scripts' are central, framing users as 'readers' of technology. Interpretative flexibility is seen as constrained 'because the design and the production of machines entail a process of configuring the user' (Woolgar, 1991, in Oudhorn and Pinch, 2008: 548), in short – the technology cannot be used in any way the users want because of the limitations built into the technology by designers. The representation of (certain kinds of) users (real or imagined) in designing and testing technology is crucial in the process of co-constructing 'not-yet-settled' technology and 'not-yet-settled' users. 'Script' (Latour, 1992) denotes how designers anticipate user patterns and behaviours, and build it into the technologies, hence enabling and constraining sociotechnical relationships. Scripts are sometimes opposed by users refusing to submit to the anticipated use. Later studies have emphasised how designers are also configured by users and their organisations, making the process of configuration a mutual matter (Mackay et al., 2000, in Oudshorn and Pinch, 2008: 549). This view challenges social constructivist approaches by giving more agency to non-humans, destabilising both users and products and viewing them both as relational effects. Actor-network approaches are criticised, however, for ending up putting 'more weight to the world of designers and technological objects' (ibid.: 551). This limited understanding of user relations, particularly economic relationships, within STS has recently led to studies of markets and marketing, where, for example, institutionalisation of market forms, shaping of distribution networks and 'branding' have been analysed as attempts to shape and control the relations between products and users.

In outlining a 'performative' view of markets, Kjellberg and Helgesson have created a conceptual model of 'markets as constituted by practice'. They suggest that markets are made of normalising practices ('to establish normative objectives'), representational practices (to depict markets), exchange practices ('to realize individual economic exchanges') (2007b: 137) and the continuous translations between them. Segmentation, for instance, is thereby viewed as an ontological act, not only of discovering and representing segments of customers, but also of participating in shaping and maintaining these groups in practice. The example of segmentation also 'illustrates how a chain of translations may allow one setting, in this case often a marketing department, to assume (or be given) the right to speak on behalf of others, in this case a number of customer groups' (Kjellberg and Helgesson, 2007b: 145). Food retailing is then used as an example of a market where 'exchange, normalization, and representational practices [are] linked through highly stabilized chains of translations', and with a clear 'division of labour' between different actors, such as food regulation authorities (normalisation), marketing agencies (representation) and retail (exchange). This makes it harder to introduce new products that cross the established boundaries between product categories and user practices, as they may require modifications in all the three kinds of market practices. My study illustrates this view of markets as ongoing and performative, showing how difficult it is to produce representations of non-existing markets, to negotiate exchange when there are few associations to the new product, and to actually produce a material product based on anticipated quality demands not supported by existing regulations (of micro-biological standards in the fish industry).

Araujo (2007) outlines marketing activities as both performing calculations in institutionally stable market structures and 'qualculations' (qualitative judgement) in the shaping and reshaping of market 'frames'. He defines the qualitative side in the following way, 'the work of qualifying goods, of imbuing them with specific qualities, is a distributed effort involving both market professionals and the final user' (Araujo, 2007: 214). Thus, he expands on Callon and associates (e.g., Callon, 1999; Akrich et al., 2002a; 2002b; Callon and Muniesa, 2005), who are viewed as over-emphasising the calculation side, and leaving little room for interaction between market participants. Within this tension 'between markets as institutions and markets as dynamic, learning spaces' (Araujo, 2007: 215), the challenge of innovation can be found: qualifying new products for established market practices by adapting them to existing categories and practices and/or establish new ones altogether (Azimont and Araujo, 2007), as 'there is no calculation without qualculation and new forms of qualculation, if successful, can destabilise existing forms of calculation and usher in new ones' (Araujo, 2007: 222). Studies in this field, however, are more concerned with the institutionalisation of markets and market practices, than with (commercial) users. Cochoy (2005) has studied how big business (food retail) has produced 'customers' as collective and segmented actors in order to gain control of globalising markets. He goes on to describe how lawmakers created 'consumers' endowed with legal rights, thereby ending up with the present hybridisation of business and law that was followed by standardisation. He argues that this has subordinated the industrial customer to the consumer customer. Araujo (2007) suggests, on the other hand, that in mass retail, the power is on the supply side, choreographing the buying process for the individual consumer, while in business markets there is a more equal distribution of power between the supply and the demand side. My case study shows how industrial interaction between suppliers and retail still seems to be more central to the shaping of food markets than the influence of consumers. Before the consumer has an option at all, the question is: will the distribution actor (retail or restaurant) offer it to their customers? Resembling Woolgar's (1991) 'configuring the user', Cochoy sees the need for educating the consumer, 'selling her not only the product, for example, but the "use that goes with it"' (2005: S38), but again, the user is not granted an active role in co-producing the 'market'.

Lury (2004) has analysed brands and how they take part in shaping the economy. She argues that the 'object-ivity of the brand emerges out of relations between its parts, or rather its products (or services), and in the organisation of a controlled relation to its environment' (ibid.: 2). The brand thereby comes to serve as an alternative coordination mechanism to price, with more 'qualitative intensity'. The case study of this book shows this difference between *price* and *brand* as coordinating mechanisms in an interesting way: it is the story of movement, or translation, of a set of raw materials and products to escape the domination of price as coordination mechanism by replacing it with the brand instead. This provides the supplier with a larger repertoire of product attributes to manipulate and stabilise the relation to users. Producing a brand, then, is 'the management of relations between [product] attributes', such as price, packaging, place and promotion (ibid.: 5). Brands are in this way both used to produce difference – for example, related to fast-changing fashions and collections of products – and sameness, by the brand acting 'as a guarantor of the consistency of quality' (ibid.: 9). This is also the case with the Salma brand embracing a high-quality strategy in a market segment dominated by anonymous (non-branded and with invisible origin) products with high variation in a number of qualitative aspects. Still, the way Lury positions the customer-user outside the set of relationships comprising the brand, making an asymmetric distinction between human and non-human actants, seems to be a bit problematic. The interaction between users and brand simultaneously takes part in shaping (changing and/or maintaining) the brand itself, hence users cannot be analysed as 'outside' the 'brand actor-network'. For a brand to exist at all, at least as a mechanism of coordination in the economy, it has to be continuously related not only to products and marketing techniques, but also to users. In this sense, branding too is a matter of co-creation, of mutual construction of producers, products, markets and users.

While insisting on the heterogeneous (social, material and economical, that is) character of market practices, both products (material resources) and users seem under-studied in this stream of research. First, there is the representation of the actual (social and material) properties of the products (or services) in question. When talking about fresh food products, it becomes clear that the products in themselves – and not only their representations as part of socially constructed 'structures' of market practices – have a great impact on the shaping of their markets and market practices. The material relations within fresh foods are fragile to the extent that they require high degrees of caution during production, packaging, transport and presentation. Everything, from micro-biological organisms through humans, to sunlight and temperatures may serve either to protect and even improve quality, or weaken and damage it. This has led, for example, to distribution innovations like cold-chain technology and supply chain management having huge success and giving great powers to the ones controlling them. Second, users suffer either from being positioned on the 'outside' of market practice or from being only present via (indirect) representations; anticipated and categorised into the 'scripts' of various marketing tools, such as segmentation and branding. I will now turn to recent research on industrial networks (Håkansson and Waluszewski, 2001a; 2007a), and in particular to how they treat these two factors of resources and users.

As mentioned previously, Håkansson and Waluszewski (2001b) observe how resources often seemed to be 'cemented' upon each other, thus being hard to change or replace; yet apparently stable resource combinations could sometimes suddenly disintegrate. In addition, intentions, whether weak or strong, could both stabilise and destabilise resource combinations (ibid.: 1). In order to describe and understand these forces, the concept of *friction*[5] was introduced. From this perspective, innovation is a challenging task: first, for creating movement at all (destabilisation) and, second, for coping with what is new without destroying what is current. It does not make it easier that change often has a number of unintended (and often unpredictable) consequences in related interfaces. Still, it is also a clear limitation of human agency, as the use and value of any resource is determined by the relationships and interaction processes in which it is embedded (ibid.: 4). Hence, effects are never just local; they become distributed through friction with other interfaces with other resources, transforming them too. This is a process of bringing two or more histories together, with an uncertain outcome and no 'best' solutions, and with a challenge of integrating the new interface with related existing interfaces (ibid.: 15).

One reason why friction also produces destabilisation effects is that simultaneous processes are connected through friction, which allows the same interface to be activated in several change processes. In this way, friction can also strengthen change (ibid.: 17). The more the focal resources are embedded in other interfaces, the more friction there is, the more resources will be affected and the more power needed to initiate change (ibid.: 18). In other words, this is an argument for embedding resources, a parallel to Law's (1994) argument of heterogeneous engineering to stabilise actor-networks. What then influences the degree of friction? Håkansson and Waluszewski (2001b) found 'economic heaviness', meaning investments in material and immaterial resources, and the combination of these in complex webs of relationships to be a conservative force. They argue that attempts to change are forced to be 'economical', from the established structure's point of view (ibid.: 23), a conservative factor in innovation. On the other hand, this seems to be challenged by Akrich et al. (2002) in their empirical example of process innovation where solutions were implemented in spite of no real cutting in costs, and no real increase in effectiveness. This is then made into an argument for how *interests and intentions* of different actors are constantly negotiated, and the outcome – whose 'forces' succeed – is left an empirical question. Friction distributes the tension of innovation to other related interfaces; combinations will be questioned and thus put under pressure to develop.

According to Håkansson and Waluszewski (2007b), knowledge of use resides in the relations between suppliers and users, as well as between resources and users. They argue that academic knowledge has a long way to go before being usable by business. While Woolgar (1991) emphasised the 'configuration of users', hence privileging designers of technology, Håkansson and Waluszewski argue that it is users which determine the value and use of innovations, and, further, they point to the dilemma of specialised versus generalised knowledge. Although the industrial logic of 'economies of scale' requires comprehensive generalisation of knowledge (assuming that many users need similar products, and will use the products in similar ways), it is often necessary to embed the innovation into specialised user practices in order to succeed with creating use. Håkansson and Waluszewski's argument for the need to embed innovations into existing economic and technical relations in order to get used implies trial-and-error learning process in interaction with users, and thus that users should be involved at an early stage in the process.

I have three questions about this that are of relevance to the discussion in this chapter. First, how can an actor involve others in something that s/he is unfamiliar with, as well as not knowing whether that something is feasible at all? We saw how the experienced company did not want or dare to contact new or established retail customers with the innovation before it had taken shape to a certain extent. They needed 'something to present'. Second, Håkansson and Waluszewski (2001a; 2007a) speak almost entirely of the 'economic world', as if this part of our reality is not also simultaneously social, professional, material, cultural and political. All these realities affect the network at the same time. Thus, I wonder if the available range of actors (and therefore users) available in the analysis is too limited? However, this is much more problematic in the 'user-driven' innovation literature, where 'users', for the most part, are synonymous with customers (more often consumers than industrial customers). Third, the argument that conservatism inherent in industrial networks (resource combinations), and the following requirement to adapt innovations closely to established practice, is powerful in pointing towards incremental innovation. However, it also seems to hide the need and ability of actor-networks to redefine (i.e., translate) existing networks into representing and realising a new set of goals and interests. Nevertheless, the mechanism is still basically the same: an innovation needs to, at all stages of its development, find allies, or users, who find they can *use* it for something, in other words, have *interest* in it.

In sum, while industrial customers are increasingly included in the analytical framework, consumer users are kept 'on the outside' of the set of relations 'internal' to the analysis in these streams of research. In addition, none of the above included 'non-customer users' as part of their framework, or, to be more precise, none gave them status as 'users'. Not that all such users are left out, but still I argue that analysing a wider array of participants in the process in terms of their *use of the innovation* will add a powerful dimension to how the innovation is connected (or not) to other processes/actor-networks during its journey towards realisation and commercialisation. It is also notable that, in these contributions, products are for the most part treated indirectly, via their representations of such elements as marketing practice, and hence lose some of their real/material impact on use, users and markets. When I argue that more kinds of users should be included, my point is that innovations will, along the way, be used for a number of different purposes in a number of different practices – for example, research groups that use the fish salami and the pre-rigor salmon for producing academic knowledge; political actors using these projects to boost their vision of 'blue-green' innovation; and raw material suppliers using the project as vehicle out of raw material markets. In this respect, the focus on configuring users and building user anticipations (privileging some kinds of use/users while restricting others) into the technology, marketing and distribution calls for a better understanding of the user side, including their 'real' needs and their response and participation in shaping markets for innovation.

Mutual translation of interests

Based on this case study and critical reviews of the literature, the fact that innovation processes are situated in networks of intertwined processes was built into the conceptual model in Chapter 5. What are the implications of this? First, it adds to the complexity of innovation processes, including increasing the number of 'nobody-knows' problems, and the number of elements to relate to each other to stabilise the network. Second, it means that we need to widen the analytic scope of innovation ethnographies to include more kinds of users. The literature is clear that innovations need to be embedded into heterogeneous networks to find use, by mobilising and adapting to other processes with other interests and aims. However, the relevant constellations in which to embed an innovation may change over time, as the innovation process moves through different phases and challenges. Innovation is a kind of 'stabilisation process', in which the stability of networks is the result of embeddedness of elements: resources, actors and practices. Movement in such embedded networks creates friction, which is a creative and destructive force, although due to the 'economic heaviness' of prior investments friction tends to work in a conservative manner, similar to path dependence, privileging continuations and incremental changes of the existing practice. It is hard to destabilise established constellations, and it is hard to create something new without destroying some of what is current. However, innovation does not always work towards 'economising'; it can also manifest itself as a struggle for (the right to) redefine networks, by negotiation, force and alliances. Friction in embedded networks also means that effects of movement are distributed throughout networks, making such processes highly uncertain. Potential constellations of actors and resources are explored via trial-and-error learning.

In the food industry, power increasingly seems to move to the big retail chains (see also the mobilisation section in this chapter), hence making it harder for producers to translate new products to *use* and *users* on their own. A few huge corporations and their portfolios of retail chains are becoming 'centres of calculation' in the food industry, with increasing ability to control both what is produced by farmers and industrial producers, and what is granted access to supermarket shelves across the globe. Globally, Busch (2004; 2007) has documented how corporations like Walmart, Royal Ahold and Carrefour are developing a rigid design of distribution based on cold-chain technology and economy of scale in their supply chain management systems. This leads to high demands for adaptation by suppliers of all kinds. Retail is moving towards distributed and centralised actor-networks with immense power, ability to act in coordination and ability to coordinate and direct the action of others – acting at a distance, and be loyally represented by a huge distribution network (Cochoy, 2005; Busch, 2007). The same kind of development started during the 1980s at a national level in Norway, with NorgesGruppen, ICA (now owned by Royal Ahold), Coop and Reitangruppen. Together they had, by 1999, reached a 99 per cent market share, from 46 per cent in 1990 (Knutsen, 2007). In this new situation, food producers have to turn from focusing on their own raw material suppliers to focusing on the distributors. Unless they associate and align themselves with these powerful retail actors, it is impossible to succeed with new products on an industrial scale. It was in this setting, both globally and nationally, that Salmon Brands sought to commercialise their salmon salami. After failing to get access to the 'ideal customer' (i.e., restaurant actors), they had to go to the retail sector. The relative stability of industrial networks, so often observed by industrial network researchers, was not easily mobilised on behalf of the salmon salami. Still, stability is relative: an element may resist, break down or disappear (recombine with another network), hence possibly shaking the whole network. Existing constellations were not easily altered to include salmon salami. This would perhaps have demanded too much by way of restructuring supermarket shelves, restaurant recipes and consumption practices (and preferences), and hence been too risky.

When it comes to marketing practices, as well as the tools and strategies producers and distributors use to create and maintain markets for their products, Callon and fellows distinguish between calculation and qualculation. From an innovation point of view, the issue is to succeed in qualifying the product for calculation – in other words, work hard on qualculation. This is because qualculation may destabilise the present state and allow for innovations by adapting to or changing existing categories, or developing new ones. The Umi No Kami/Salma case provides an example of the challenges of qualculation, or of qualifying a new product for established market and distribution practices. The distribution of power is shifting, in various industries, towards the consumers according to Cochoy (2005), and towards retailers according to Araujo (2007). Anyhow, power in these settings depends on the number of allies of the various constellations (actor-networks), and presently in the food industry there seems to me to be a shift from producers to distributors, while keeping some power with the consumers as far as they have choice, and are represented by market research and law.

The interconnecting of the innovation and its development process to other processes in the framework clearly shows a distinction between creating new relationships, on the one hand, and utilising and redefining established relationships, on the other. I have previously argued that the innovations that are most similar to and/or most adapted to existing practices are more likely to succeed, due to the inherent development risk in any innovation process. In commercial innovation, businesses are forced to minimise development costs via radical simplification and adaptation of what already exists. Historically, businesses have not been the proponents of radical innovation. Instead, publicly financed projects, programmes and regulations have often driven such processes. In strictly commercial innovation processes, therefore, one would expect as little market development as possible. This conservatism, advocated by industrial network researchers (Håkansson and Waluszewski, 2001a; 2007a) carries a critique of the actor perspective in business studies, as well as a bias towards 'designers' of innovation in STS perspectives such as actor-network theory. By this logic, Tine should probably never have become involved in a project as radical as the fish salami project, due to the degree of uncertainty, the development risk, the lack of commercial users, etc. So why did they still do it? In this case study, there were some specific reasons, in addition to commerce, for Tine to go for this project, as well as the other biomarine projects. It was used as part of the strategic process aimed at repositioning the corporation from dairy/agro-food to 'food corporation'. In addition, it was used in national discourses of food/industrial politics and rural politics, showing that Tine 'took responsibility' within biomarine activities, possibly strengthening their position and goodwill with governmental ministries. Lastly, there was a clear aspect of epistemic, or professional, interest in the project, fuelling it with enthusiasm, as well as providing the basis for external financing from the Research Council.

However, the challenge goes the other way too. The conservatism of industrial networks ('economic heaviness' and friction) is challenged by actor-network theory's conception of 'translation' (mobilisation and redefinition). In order to create *use* for innovations, other actor-networks have to be mobilised and redefined. It is clear that networks are never created from scratch, but there is (sometimes) room for renegotiating the interests, programmes and constellations of networks. Further, in order to succeed with innovation processes, it is often a battle of either translating others or being translated, hence there is the idea that innovation requires negotiating and compromising in complex networks, and not leaving any of the involved parties unmarked. It is still clear from this that utilising and redefining established relationships may be easier than establishing new ones. This was also observed in the case study; Salmon Brands had problems establishing new relationships on distribution, both in the retail and restaurant industries. When they did get access, it was by redefining well-established relationships with retailers, first in Germany and then (with Salma Fresh) in Norway. On the supply and production side, the new relationship between Tine and Bremnes Seashore came about by the active brokering of individuals with close relationships to both companies. Moreover, it happened after Bremnes had tried to redefine their relationships to their own distribution partners without success, and thus they had a strong motivation to find new paths to get return on their innovation investments. It is a challenge to establish relationships, and to recruit allies; these potential allies have to be able to make use of the situation for their own purposes. I argue that new user–producer relationships are hard to establish during innovation, and they will be fragile due to a lack of tangled interests and resources, providing little commitment from the (new) user. Therefore, the lack of 'first connections' in many innovation processes may prove to be critical to the outcome, thereby demonstrating a clear advantage of mobilising established relations, if they are available. MRC and Fairprice in Singapore, as well as Nutrimer in France, could just drop the salmon salami project without warning, without giving any reason and without risking much.

Furthermore, in the 'outwards' organising of the project, Tine and Salmon Brands were very conscious about the need to include the consumer in their product development, researching various consumers' environments and asking samples of consumers about their responses and potential use of a fish salami. But, as a result of their reluctance to involve distributors early in the process, they failed to get answers to the crucial question: within whose product range would the innovation be a good fit? This is also a challenge to the end-user/consumer bias within the literature on user-driven innovation. Who are the industrial users of such a product, what would they be interested in using it for and what shape should it therefore have? Before the consumers have a choice at all, the product will need to be strongly aligned with committed distribution partners (in this case retail or restaurants). Tine knew about this challenge, but did not want to risk making a fool of itself by presenting the idea before it 'had something to present' – that is, before it had a clearer conception of the product and a stable prototype. Further, when it started the recruitment of customers, the target users were hardly specified; it was a trial-and-error learning process that, in this case, led nowhere. This raises pertinent questions. When should (commercial) partners be involved, how far should the innovation be developed before exposing it to the risk of a bad response from potential allies and how much simplification (removing ambiguity/multiplicity) is needed to get the message across? On the other hand, the innovation should not be too stable, to allow for the necessary 'interpretative flexibility' so that the innovation may be adapted to the interests and use of potential partners. This leads to another important aspect of users and innovation, namely that attempts at building the innovation into commercial relations are likely to destabilise it and produce new phases of development, of mobilisation and exploration. In the Umi No Kami/Salma case, we saw how conceptions of users and markets were built into the product and its network, and how a range of potential versions and directions was maintained simultaneously. These propositions about users will rarely fully match up, because finding or creating *use* for the innovation in other actor-networks means that the innovation needs *renegotiation* and, therefore, that it often will be thrown back into new rounds of development (mobilisation and exploration). With regard to the 'internal' organising of development (here, in organising relations to users), confrontations between mobilisation and exploration are often destructive, thus actors seek to avoid involving themselves in more relations than necessary. Simplifications of networks might be necessary, and the confrontations that the actor-network goes for are those they think they can win. While sometimes necessary, this reluctance to interact may again lead to sub-optimal mobilisation or exploration. Hence, I argue that when partially stabilised innovations (and their internal propositions about users) are tested with potential users, new propositions and adaptations of the established will arise, and thus lead to new development phases and new selection processes.

A basic logic of this innovation process can be seen as 'connect and stabilise', whether regarding material combinations or social relations. The interaction and adaptation of each human and non-human element to be included was, however, not straightforward. The mutuality of technology and use continued and perhaps even intensified during the process of interacting with potential users. If the process of connecting and stabilising technical and conceptual relations was hard work with uncertain outcomes, the process of connecting and stabilising relations between Salma and users was no less demanding. Partly stable aspects of Salma had to be destabilised and changed several times in relation to responses and demands from several sceptical and demanding users.

As previously mentioned, this case study exemplifies the connection of past and present through friction, although it does not entirely work in the conservative way suggested by Håkansson and Waluszewski (2001a; 2001b). Compared to Tine's core business, this project was of a relatively small size, and its failure would not threaten Tine's resources in any significant way. Thus, friction here had a less conservative effect, because the risk related to already established activities was low, and current investments and technological systems would continue independent of the biomarine projects. As for the other party, the fish farm, things were rather different. Relative to its size, its investments into this joint venture made it more dependent on the outcome of this project.

The ideal of marketing practice is increasingly moving towards interacting and relating with users (whether industrial customers or end consumers), and is also becoming the norm for innovation practice (e.g., Von Hippel, 2005). However, when dealing with radical innovation (where 'radical' means unfamiliar, boundary-crossing and discontinuous), user-involvement brings with it some challenges regarding methods, timing and strategy. In this case study, when Tine R&D thought they were almost finished with a product, and eventually presented it to potential customers, the product had typically to be sent back to the lab again for further development, improvement or changes. The question, then, is whether this extra work could have been avoided if users had been involved earlier in the process, or whether this would only have distracted from the fragile process of technology development. Integration of development and marketing is thus depicted as both necessary for understanding and meeting the user demands, and difficult for the lack of methods for involving users before having produced 'something concrete' to represent the project.

The users who were 'at hand', however, were chefs. They were very much involved in the project, at different stages of the development. With creativity and skill way above that of the average consumer, they took on the task of situating the product within familiar eating practices, demonstrating possible dishes and use situations, associating the fish salami (and later the fresh loins) with other ingredients within a set of popular and exclusive food traditions. Thus, the emergent object, neither technologically nor conceptually stable, got some help from a few 'lead-users', and a dedicated and enthusiastic project team of researchers, product developers and marketers. Even if this resulted in wonderful presentations and representations of the product, and thus contributed to shaping the brand image of gourmet quality and use, it is interesting to question whether top-class chefs were the *right* lead-users to align with. They were, no doubt, of high competence and creativity, and added a portion of gourmet credibility to the Salma brand. Yet, on the other hand, chefs and private consumers are different *kinds* of users. The use of the product is likely to be different between a chef in a restaurant and a consumer buying something for his/her family meals.

Their competences are different, as are their preferences of taste and use situations. This may, however, have been more problematic with the Cured version than with Salma Fresh. While Fresh *may be* used in some of the suggested recipes, it may also be used in the same way people have traditionally prepared their salmon, with the added benefits of having had the skin and bones removed.

Yet, were all *necessary* associations in place? Would it have been possible to work together with industrial users from the start of the salami project (UNK)? Would Tine have found actors interested in exploring opportunities from fermenting fish together with them? Finally, would this have led to a successful commercialisation of the fish salami? No one knows. Although it did not involve industrial users early in the product development process on Salma Fresh, the finished product and concept turned out to be of great interest to both retail chains and (individual) restaurants. Established relationships between Tine and Norwegian retail chains could also be used for selling fish.

An aspect that brought the pragmatics of the commercialisation process to test was the difference between the anticipated user – as s/he was built into the product and the brand concept – and the actual (and potential) users that the Salma team interacted with in practice. The (first) anticipated users were curious, creative and high-paying users of novel and healthy products, either in private households or in restaurant kitchens, able to translate a salmon salami into their patterns of cooking. The (potential and actual) users interacting with Kiland and Swendsen on Salma Cured in practice were curious indeed, but also sceptical and seriously concerned with problematic issues like the associations of salami with meat, price and packaging. MRC even demanded a larger change with regard to the use and technical characteristics. Tensions appeared between the anticipated and actual (potential) users that could not necessarily be easily resolved. The shift towards Salma Fresh, or selling fresh loins under the same brand, was consistent with this pragmatic drive towards selling something to someone, hence going for the easiest alternative.

Tine's user relations and marketing expertise could not be used to bring Salma Cured all the way to supermarket shelves, or in any other way towards the shopping baskets and food plates of consumers. For this, they had to simplify the product (paradoxically coming up with a product perceived as being more exclusive), and associate it closer with established categories, practices and networks of food.

7
Conclusion

I am ambivalent towards writing conclusions. To wrap up years of research and rich descriptions of complex processes in just a few pages provokes a fear of reductionism and of restricting the book too much. Nevertheless, I will in this short chapter compress the content from the previous chapters to provide a summary of the contributions to innovation process research, and point out a few areas of further research.

This study contributes by unpacking the black-box of innovation processes, particularly related to controversies and friction, through its investigation of innovation processes and practices in one small sector of the business world: a Norwegian agricultural cooperative, Tine, and its counterparts in the agricultural and biomarine industries. It is an ethnographic case study of the organising of innovation processes; the development and commercialisation of hybrid technologies and products between aquaculture and agriculture. What I have described is the emergence of a possibility: the possibility of industrialising fish, and some early attempts at doing so. The ongoing micro-practices of realising the concept that came to be called Salma should be viewed as a mutually constitutive part of what might become new industrial practices between agri- and aquaculture. It illustrates how a changing 'macro-structure' creates new opportunities, but also a bottom-up perspective of how new industrial and market practices are made.

From an overall interest in how innovation processes evolve over time, I aimed to understand more of the organising of innovation processes crossing technological, organisational and industrial boundaries in time. Moreover, I wished to increase our understanding of efforts to realise and stabilise new knowledge and technology in innovative ventures, with particular emphasis on how controversy is overcome as part of the process of stabilising relations. My hope was that this could contribute to innovation studies by shedding some light on the question of how knowledge, which is developed in certain contexts and embodied in certain objects and practices, is rendered mobile, translated, combined and made stable in new settings, objects and practices underpinning innovative ventures.

Innovation in the setting of my case study implied the development and commercialisation of knowledge and technology. This was manifested in new products and ingredients, and in new practices that crossed, or reorganised, the traditional boundaries between agri- and aqua-culture. A dual dynamic came to the fore in the analysis of the case, namely between mobilising actor-networks on the one hand and exploring knowledge on the other. These 'sub-processes' of the larger innovation process sometimes drew on each other; at other times they did not interact at all; and sometimes they came to confront each other – with potentially serious implications for the future of the innovation. This dual dynamic became an 'interpretative scheme' from which I have structured a theorising discussion on innovation processes.

Innovation: processes, controversies and networks

Empirically, this book has provided rich insights on a complex phenomenon: industrial innovation crossing multiple boundaries. It has also given a relatively detailed description of a project that may or may not come to be representative of convergence between aquaculture and agriculture. In the contrasting of these two industries some of the distinctive characteristics of their practices have been revealed: the international and raw –material-oriented fish industry; and the national and industrialised dairy industry. The case study shows that industrial interaction between suppliers and retailers still seems to be more central to the shaping of (Norwegian) food markets than the influence of consumers. However, in this new situation of shifting power from producers to distributors, food producers have to turn from focusing on their own raw material suppliers to focusing on the distributors. In this situation, the 'next' users in the value chain need to be included in the innovation process, and the crucial question to ask is: within whose product range would the innovation be a good fit?

The case study, written as a 'thick description', may be read from different theoretical perspectives: as a story of learning, industrial change, strategising or innovation, to mention a few. By constructing and amplifying the distinction between mobilisation and exploration in a bipolar model, I explain some of the micro-dynamics of innovation processes

from an angle that, to the best of my knowledge, has not yet been sufficiently described in the literature. First, during (sub-) processes of
mobilisation, actor-networks are recruited and committed to things
with which they are initially unfamiliar: an idea, a prospect or a prototype of something that may or may not become feasible and useable.
Yet, to enable mobilisation, a degree of certainty has to be presumed.
Second, during (sub-)processes of *knowledge exploration*, the aim is to create knowledge – to explore the object and its potentials – and therefore
change is unavoidable. Moreover, this process of generating knowledge
tends to multiply (alternatives of) the object, and hence increase rather
than decrease the uncertainty/complexity – or development risk – of the
project. Mobilisation and exploration are contrary forces in this model,
and sometimes it appears almost like the innovation process is at war
with itself. Whereas mobilisation is directed towards aligning interests
and reducing risk, exploration is directed towards formulating and testing propositions about reality. While mobilisation seeks to converge,
exploration frequently leads to divergence of the innovation. Finally,
the *interaction* between mobilisation and exploration processes on the
one hand, and between different actor-networks/organising processes
on the other, often leads to controversies and compromises that may set
the project off in new directions.

In relation to ANT/STS, the study shows how the pragmatics of business (which may be understood as a set of practices connected to economic theory, consumer practices, industrial networks, marketing/
branding practices, etc.) may compromise the technological passions
and interests driving technical innovation, making science less 'pure',
less hegemonic and less seducing. In the meeting between inventions
of technoscience and the tough 'realities' of business, it is an open question what remains of the initial innovation, how it is combined with
existing business practices in order to find use(-rs) and how it takes
part in reconfiguring and reconstituting those practices. In relation to
innovation management and industrial networks, the study shows how
technoscience, and its creative chaos (of ideas and direction) and rigidity (of method), sometimes serves as precondition, one of a number of
resources, for creating new commercial practices.

The presence of a number of uncertainties – 'nobody-knows' problems – frequently produces high development risk in innovation projects.
Although being experts in their respective fields, I argue that innovators' lack of knowledge pertains to the connection and translation of
knowledge and technology between settings. Putting knowledge, and
the lack thereof, at the centre of attention, the framework suggests that

innovators have to produce two different kinds of knowledge: first, a chain of arguments suited for convincing, mobilising, maintaining and removing (parts of) actor-networks and their resources; second, they need to produce testable propositions about reality (e.g., of how to make the technology work and what users have interest in such a product). Innovation processes are propositional at their core. The original idea is a proposition about the potential that stems from a new combination of elements. This idea needs some resources to get started, and then the idea needs to be explored in practice – testing whether and how the proposition may hold. This will normally happen by breaking the original idea into a series of new and 'smaller' propositions; as the innovation is opened up and investigated, it is revealed as a more or less complex set of problems, all having many different solutions *in potentia*. However, in order to enrol allies, it is necessary to make the idea and concept converge on a number of aspects, and this will often create a 'lock-in' for the subsequent process – at least for a period of time. I further argue that every 'nobody-knows' problem of an innovation has to be taken through several stages to be resolved. Further, the complexity – and hence risk – of innovation processes is equal to the number of 'nobody-knows' problems times the number of stages they have to be taken through.

Mobilising actor-networks

In this book I argue that power to mobilise elements and decisions for innovation is produced by constructing a (more or less coherent) chain of arguments. In other words, mobilisation of actor-networks is based on a relational logic of 'power production' – that is, of carefully building (or connecting to) networks of human and non-human elements with interests in realising the innovation. Hence, constructing an actor-network around a new idea is an exercise in creative connecting of actors, networks, resources and ideas. In this sense, this part of the process more or less ignores the 'truth'; rather, it is about producing power effects – that is, mobilising actors and resources on behalf of the innovation, and translating their interests into a common project. Still, if and when an actor-network is mobilised, the elements employed in the chain of arguments may produce frames and evaluation principles that define the room for action in the project. Hence, temporal lock-ins may be enforced that cannot easily be broken out of in the subsequent parts of the innovation process. In Chapter 6 I have derived some more specific suggestions for theorising the mobilisation aspect of innovation processes.

I argue that there is a relationship between mobilisation and time. In the case study, elements enrolled and mobilised in the innovation process tended to have more influence on the innovation the later in the process they were included. The reason for this effect seems two-fold. First, the element representing the most recent changes – improvements or alterations – has more narrative power in representations of the innovation process, such as project reports and presentations, in their symbolic power as representatives of 'progress' in processes of high uncertainty and unclear evaluation criteria. Second, in order to mobilise new actors and resources into the process, sometimes the significance of the results already achieved has to be downplayed in order to increase the necessity of the new. Argumentative power may be produced by increasing the importance of the new relative to the old.

I also argue for the *exploitation of uncertainty* in interaction between (potential) allies in efforts to stage and mobilise resources for innovation. The existence of knowledge 'fields' and 'practices', as well as 'frames of reference', means that different actors have different experiences and expertise related to the characteristics (and potential) of the elements recombined into a new idea. Moreover, different actors are situated within different sets of relationships. However, few – if any – know what it takes to relate previously unrelated elements to each other. I argue that there are three aspects of the relationship between lack of knowledge and the mobilisation of actors and resources for realising innovative ideas (the building of actor-networks) that need to be explicated. First, asymmetrical knowledge and experience may be used during interaction to exploit the other. In the process of building arguments to convince others on behalf of an innovation, presumptive competent actors are mobilised to represent the innovation as something worth pursuing. Second, manipulation based on asymmetrical knowledge consists of mobilising apparent authority. Hence, if more 'radical' ideas will be impossible to evaluate in objective terms, those with more experience with (some of) the elements involved might be able to exploit actors with less or other types of experience. Third, asymmetrical knowledge and mobilised authority may therefore contribute to the construction of a 'market for ideas' by those who have 'knowledge advantages' – that is, work as a trigger for economic exchange of ideas. In any case, choices regarding innovative ideas always have to be made based on limited knowledge.

Exploring knowledge

Uncertainty, or the lack of knowledge, in innovation is a problem calling for 'exploration'; an active learning process. Exploration is the process of actually testing and developing the innovation. Such exploration processes are typically about imagining and formulating testable solutions – for example, in the form of theories or propositions for problems of a technical, social and/or economic character – and then testing these solutions in practice. The exploration process, of knowledge generation through proposing, testing, interacting and adapting (aspects of) the innovation, is supposed to move asymptotically towards solid knowledge. The paradox is that while the aim of exploration processes is to produce knowledge, they (almost) always produce complexity: multiplicity, ambiguity, choice. Even if sometimes the knowledge-generating process succeeds in providing clear and singular answers, most of the time the object and its complex of (potential) relations in this case study expanded during exploration. I argue that this partly has to do with the innovation, at least in its early phases, being unstable within several dimensions, and that its stabilisation requires exploration of a number of interconnected issues. In addition, when experts start investigating an idea, making it into an 'epistemic object' (Knorr Cetina, 2001), it opens up and becomes a complex of interesting problems and opportunities. Paradoxically, in industrial settings exploration is a process aiming for expanding and generalising the concept, often involving hypotheses about appropriation and economies of scale. The innovation has to be brought towards stabilisation as a general concept, thus facilitating scaling up. This presupposes that the concept is tested in practice; that it shows that it holds. From this I suggest some theoretical implications of the knowledge exploration (sub-) process.

A main point in my analytic scheme is that processes of mobilisation and exploration do not always interact. Sometimes they are not coupled at all, and when they do interact, tensions, confrontations and conflicts are often produced. I suggest that the reason for this is that the exploration process, of generating knowledge, challenges the mobilisation process that is already established, and creates a dynamic that may either strengthen the emerging path, or undermine and destroy it, or even lead to the establishment of new path-creation processes entirely. Often, learning leads to departing from the original idea, which may create a mismatch to the extent that a battle for the future direction is unavoidable. Therefore, interaction avoidance seems to be a common challenge of innovation processes, basically stemming from the need to handle and reduce the divergent and expanding aspect of exploration. I argue that actors tend to avoid interacting with others during exploration processes because of the risk of being influenced, moreover, that the battles that innovators take on are only the ones they think they can win. We also saw in the case study that, after a successful mobilisa-tion, the resulting framework is kept tight until new confrontations and reconfigurations enforce or enable renegotiation, either of the frame-work conditions or of the actual innovation.

While acknowledging that knowledge is situated, and hence diffi-cult to transfer across settings (Orlikowski, 2002), I argue that it is still relevant *who* is trying to move and translate knowledge. It seems that some actors are more likely to succeed than others, depending on the – sometimes unexplored – compatibilities between the specific knowl-edge regimes represented. In the case study, for its own knowledge to become translatable, Tine needed more knowledge about the practice of 'the others', and therefore it started recruiting allies on the fish side.

Network of interconnected processes

Not only do (sub-)processes of innovation (sometimes) interact, the innovation process also interacts with a number of other actor-networks in a network of interconnected processes, thereby considerably increas-ing the complexity and uncertainty. The industrial network approach (IMP) was brought in to complement actor-network theory on this part. I suggest that, from a process perspective, path dependence is better seen as relatively slower processes, maintained via carefully and often long-term assembled and intertwined networks of heterogeneous ele-ments. Movement in such embedded networks creates friction, which is a creative and destructive force, privileging continuations and incre-mental changes of the existing practice. This view of path dependence explains some of the slowness and some of the unexpected outcomes of innovation processes: (1) why innovation processes tend to take signifi-cantly more time than expected, and (2) why 'successful' innovations often are realised as incremental changes or additions to the existing set of relations. However, the conservatism of industrial networks is challenged by actor-network theory's conception of 'translation' (mobi-lise and redefine). In order to create *use* for innovations, other actor-networks have to be mobilised and redefined. It is clear that networks are never created from scratch, but there is (sometimes) room for rene-gotiating the interests, programmes and constellations of networks. Anyway, there is a distinction between creating new relationships on the one hand, and utilising and redefining established relationships on the other, and I argue that new user–producer relationships are hard to establish during innovation, and that they will be fragile due to a lack of tangled interests and resources, providing little commitment from the (new) user.

In particular, I problematise conceptions of 'users' and 'markets' for innovations, and argue that more kinds of users than just end-users should be included in the analysis of innovation processes, because innovations will, along the way towards realisation and commercialisation, be used for a number of different purposes in a number of different practices. In fact, for the stabilisation of the innovation and its actor-network in cases of radical innovation, one needs to achieve a certain degree of stability of the innovation before involving users. Otherwise, the concept will not be able to resist pressure from sceptical actors. Moreover, with a high number of unstable and variable elements in the concept, interaction with users will produce a number of new possibilities that are impossible to handle in practice. Therefore, when finally going for something, one tends to go for the alternative with the fewest (new) elements, the most similar alternative to what exists from before.

Further, building the innovation into commercial relations is likely to destabilise it and produce new phases of development, of mobilisation and exploration. Hence, finding or creating *use* for the innovation in other actor-networks means that the innovation needs renegotiation. As mentioned previously, confrontations between mobilisation and exploration are often destructive, thus actors seek to avoid involving themselves in more relations than necessary, and simplifications of networks might be necessary. And, while sometimes necessary, this reluctance to interact may again lead to sub-optimal mobilisation or exploration. I argue that when partially stabilised innovations (and their internal propositions about users) are tested with potential users, new propositions and adaptations of the established will arise, and thus lead to new development phases and new selection processes. In order to minimise such challenges, thereby reducing development costs, businesses are forced into radical simplification of the innovation and its network, and adaptation to what already exists.

In sum, I maintain that my study has contributed to our understanding of industrial innovation processes by challenging and complementing perspectives of punctuated learning (Van de Ven et al., 1999), path creation (Garud and Karnøe, 2001), market making (Kjellberg and Helgesson, 2007a, b; Araujo, 2007) and user–producer interaction (Håkansson and Waluszewski, 2007a; Oudshoorn and Pinch, 2008). My analytic scheme and subsequent theorising is consistent with the methodology of actor-network theory, while also drawing on the insights of the mentioned perspectives. However, it differs from many actor-network theory accounts and conceptions in its attempt at handling industrial innovation, rather than science and technology development. It also differs from the related emerging sociology of finance (Callon, 1999; Knorr Cetina and Preda, 2005) in dealing with 'less pure' settings. I have emphasised the controversies of innovation, both within the process itself and between the innovation process and its related network of interconnected processes.

Innovation process research

While working from a relational and process perspective, as found particularly in actor-network theory, the movement (translation) of ANT from science and technology studies to studies of industrial innovation demands developing new tools and concepts. I have engaged with the emerging conceptualisation of industrial innovation processes by developing and using an analytic scheme, including a set of process-based concepts, which I argue to be more in line with the empirical setting of industry than what was available in ANT. This was partly derived from the dynamics of the case study, and partly developed from drawing on insights and concepts from innovation management and the industrial network approach (IMP).

The connecting of ANT and IMP was suggested by Mattsson (2003) and, on the issue of market making/marketing, this has already produced interesting insights (e.g., Araujo, 1998; 2007; Kjellberg and Helgesson, 2006). My study adds to this by following an innovation process from idea, through technology development and conceptualisation, to commercialisation. Starting out from an ANT perspective, in my view, it is possible – and rewarding – to combine some aspects of IMP with ANT. In particular, the 'resource perspective' in IMP has some clear similarities with the basic thinking of ANT. Resources in IMP are heterogeneous, relational and are seen to interact, meaning that resources only get their value and identity from their relations. If including humans, groups and organisations, this is very similar to ANT's view of actors and actor-networks. Therefore, I could import in particular two aspects from IMP to my analysis. First, the concept of friction helps me explain the problems of associating something new within, or changing, established practices. Second, the stringent focus on (relatively stable) relationships in industrial networks, giving an 'outsider' view on innovation processes, has provoked a stronger interpretation of emerging actor-networks' interconnectedness to other processes and other actor-networks. As far as I can understand, however, it seems more difficult to combine the activity and the actor perspectives of IMP with ANT. While perhaps not being *necessary*, activities and actors in IMP tend to be depicted in more structural – and thus less interactive/process-based – terms.

I have further explored in practice the possibilities and the challenges of doing ethnographic case studies of innovation processes. On the one hand, it demands open access to relevant organisational settings, and brings with it challenges of observing in multiple places and of not knowing when and where the important events happen. On the other hand, it provides opportunities to produce a more fine-grained understanding of innovation processes than that which has been produced by much of the innovation process literature, as demonstrated, for example, in the identification of some of the drivers for divergence and convergence during innovation.

Afterword

There has been a call for studies of innovation processes in organisation and management research for the last 25 years, yet our understanding of innovation processes is still limited. Although researchers like Kline and Rosenberg (1986) and Van de Ven et al. (1999) have pioneered the field through empirically and analytically rigorous studies, the topic is still far from illuminated in all its complexity. Similarly there have been recent calls for process studies of organisation (e.g., Tsoukas and Chia, 2002; Van de Ven and Poole, 2005), even though researchers like Weick (1979) radically challenged the field with his process-based view 30 years ago. I suggest that ethnographic and comparative ethnographic case studies represent a great potential for contributing to these fields of research.

First, there is a need for testing and tuning process-based models, like the one presented in this book, in more settings. It could, for example, be tested in studies of innovation processes on more 'strategic levels' of organisations. Another route would be to use the model for research within other industries, enabling testing of the relevance of the model as well as comparison of innovation processes across industries. Furthermore, the relevance of the model in settings of (1) service innovation and (2) entrepreneurship has not been discussed in this study; I would be curious about the differences that would appear if this was done.

Second, I think that the insights produced from my analysis of this case study are not exhaustive. At this level of detail in the empirical descriptions, more could be gained than I have been able to do in this project. In particular, I suggest that it would be interesting to identify more of the various strategies used for coping with the controversy and interaction (1) between the mobilising of actor-networks and the exploration of knowledge and (2) between the innovation process and the network of interconnected processes in which it is situated.

Third, I suggest that much could be gained from doing more systematic and comparative ethnography. Hess (1992: 16) lists how comparative ethnography has been used to study 'implicit values and meanings' within science and technology studies, particularly to reveal differences between geographical/cultural settings, discursive domains, researchers' practices and different categories of users of technologies. While many of the studies he refers to are studies of innovation, the focus has been more on the cultures and practices of scientists and technologists, than on the characteristics of the innovation process itself. Moreover, Hess refers to ethnography of science and technology rather than ethnography of industry. From ethnographic research understanding can be built bottom-up – from micro-events to a web of interacting actors and processes.

Notes

1 Understanding innovation as process

1. See Bakken and Hernes (2006) for a discussion of organising as 'both a verb and a noun', from a relational and process-based perspective.
2. The International Marketing and Purchasing Group (IMP) (www.impgroup. org).
3. The concept is borrowed from Bijker et al. (1987).
4. The International Marketing and Purchasing Group, see also www.impgroup. org.
5. For a reflexive account of ANT's past and (potential) future, see Law and Hassard (1999). Latour's (1987) introduction to ANT as *methodology* is still useful for understanding its basic principles and thinking. For a broad and somewhat eclectic collection of texts showcasing the state of the art within Science and Technology Studies, see Hackett et al. (2008), where also the 'economic turn' is represented.
6. See Mørk and Hoholm (2008) for a review and discussion of boundary crossing and boundary organising practices.
7. Similarly, Brown and Duguid (1991) argued that learning should be viewed as the bridge between working and innovating.
8. See, e.g., Garsten and de Montoya (2004) for a review and a collection of works.
9. See also Håkansson and Waluszewski (2007a) for a more economy/industry-oriented parallel based on the IMP approach.
10. Semiotic approaches are concerned with the construction and structuration of meaning, based on a relational view of language. ANT has expanded this to include also socio-material relations, thus emphasising the heterogeneous character of the social (Law, 1992; 1999).
11. I use the term 'boundary' not as a static term referring to formal organisations, or anything similar. Instead, and in line with literature both within STS (e.g., Star and Griesmer, 1989), practice-based studies of organisation (e.g., Gherardi and Nicolini, 2002) and IMP (e.g., Araujo, 1998), boundaries are viewed as being empirical phenomena temporally unfolding in practice. Boundaries may appear in many forms, and are more or less stable outcomes of 'boundary work' (e.g., Grint and Woolgar, 1997); i.e., efforts to define who is on the inside and outside of a practice.

2 Constructing ethnography

1. See also Olsen (forthcoming) for a constructive critique of such calls for 'process studies', which argues that *ontological* process metaphysics will be of limited value to the field, but that its *epistemology* will make

interesting and potentially useful contributions at the level of analytical conceptualisation.

2. It must be noted that I have not had any obligations or objections from the sponsors of the centre regarding research questions or design, other than a general expectation of 'relevance'. On the contrary, this position has granted me open and generous access to do fieldwork according to my research interests in these organisations.

3. See Chapters 1 and 6 for the presentation and discussion of some recent contributions in an emerging field of economic and industrial sociology with roots in STS.

4. There was a similar group in Tine R&D at Voll (Rogaland) too, and I also had a visit there in my early phase of getting to know the Tine organisation.

5. There is one exception to this: when referring to discussions in board meetings (based on Huse and Schøning's (2005) study), I have been asked by Tine BA – on a principled basis – to anonymise the participants and give descriptions on an aggregated level.

6. A great deal could be said about this software for qualitative research, including its basis in grounded theory, its ability to incorporate different formats and sources, and its functions of analysing and modelling qualitative data. However, as, in this thesis, I have only used it as a tool for cutting and pasting textual materials; the other features have not been used. It was very useful that I could always keep track of the exact sources of the quotes and other text fragments found in my themes and events.

3 Introducing the case study

1. The idea that there are potential synergies between agricultural R&D and aquacultural/biomarine R&D; see later in this chapter for a more detailed description.

2. Whey is a by-product from the production of white cheese; only a small portion was utilised within the food industry, and the rest is unprofitably sold as animal feed. Hence, Tine is constantly seeking new opportunities for economising on this 'idle resource'. Whey consists mainly of carbohydrates and proteins, and in this case it was the proteins – and milk proteins in general – that were under investigation.

3. 'Umi No Kami' means 'god of the seas' in Japanese, a follow-up on 'Neptun' in the first project, referring to the god of the seas in Roman mythology.

4. 'Pre-rigor' processing means processing the fish before it becomes 'death stiff' (rigor mortis), thereby, getting very fresh fillets of extraordinarily high quality. Rigor occurs just a few hours after slaughter, and it is not possible to take away skin and bones industrially during this phase. Therefore, all fish to be processed are stored for around three days before processing, according to the common procedure. This storing can also be done on a trailer on its way to Denmark or France, hence there seems to be less advantage in post-rigor processing in Norway.

5. Referring to the double meaning of 'fish stock' and 'fish power'.

6. 'NorgesGruppen is Norway's largest trading enterprise. The group's core business is grocery retailing and wholesaling. Through its chains, the group

holds a market share of 39.2 per cent of the grocery market.... A total of 1,919 grocery stores and 790 kiosks are affiliated to NorgesGruppen' (downloaded 12 May 2009 from http://www.norgesgruppen.no/norgesgruppen/norgesgruppen/english/).

7. I am largely relying on Espeli et al.'s (2006) history of the Norwegian dairy cooperative on these matters. All quotes are translated from Norwegian by me.

8. See Risan (2003) for an excellent study of the mixing of technoscience, culture and nature resulting in the 'Norwegian Red Cow' (NRF), dominating the Norwegian livestock.

9. The other research group that began systematic research on salmon was related to the Marine Research Institute, with Dag Møller as a leading figure. According to Professor Erik Slinde, their research facilities and programmes were established at about the same time, but there was no interaction between these two groups, partly due to a conflict between Harald Skjærvold and Dag Møller. Both these groups are still central in the research and development of salmon breeding.

10. From 1989 to 2004, the consumption of milk was reduced by around 33 per cent, meaning 55 litres per capita per year on average, after which it has stabilised at around 110 litres per capita (Hanne Refsholt, CEO at Tine, interview 2005).

11. The overarching strategy work in Tine has been organised into five-year cycles, where heavy analysis work has led to plans stretching five years into the future. 'TINE 2000' was made in 1994/5 and 'TINE 2005' was made in 1999/2000, etc.

12. The brand 'TINE' was first introduced for product branding in 1992 (Espeli et al., 2006: 213), but the cooperative changed name in relation to the 'corporatisation' in 2002.

13. Hanne Refsholt (director Tine R&D, later CEO at Tine), Per Magnus Mæhle (corporate director, business development at Tine), Jan Ove Tryggestad (board of directors at Tine and several of the biomarine companies) and Kåre Markussen (managing director of Tine Region North) were among the central participants in outlining the biomarine strategy on the corporate level at Tine from 1999 onwards.

14. Huse and Schøning (2005) observed Tine's board meetings and interviewed the board members during 2004. The restructuration of Tine Biomarin and its portfolio of projects, including the partnering with Bremnes Seashore, was discussed and decided by the board during this time period.

15. The list is not complete. Tine did also invest in some other companies, e.g., the biotech company Aminotech AS (later pulled out) and the fish product company Lofotprodukt (still associated).

16. According to the 24 February 2005 issue of the business newspaper *Dagens Næringsliv*, Tine had, in total, at that point in time spent more than 100 million NOK on their blue-green projects without any returns so far.

17. Based on Huse and Schøning's observations and interviews in their study of the Tine board of directors during 2004, in addition to my own conversations and interviews with participants.

18. I went to Maritex for a few days, together with project managers from Tine Ingredients, followed a set of meetings and got to speak with the

management group there. I also discussed Maritex with several people at Tine's management and at Tine R&D during my fieldwork.

19. FDA is the US Food and Drug Administration, and it has a crucial position in approving new food and medicine products. GRAS is the US approval of food safety, 'Generally Recognized As Safe'.

20. Interview in FISK Industri and Marked, 15 August 2006. Downloaded from http://www.netfisk.no/default.asp?page=3&article=292 on 18 October, 11.05.

4 Fermenting fish: innovation in practice

1. The 'Renew-programme' is an industrial research programme constituted by a collaboration between the Research Council of Norway and Innovation Norway.

2. 'Gravlax': during the Middle Ages, gravlax was made by fishermen by salting the salmon and lightly fermenting it by burying it in the sand above the high-tide line. Today, fermentation is no longer used in the production process. Instead, the salmon is 'buried' in a dry marinade of salt, sugar and dill, and cured for a few days. (Source: *Wikipedia*, retrieved 28 July 2007, 13.56, UTC).

3. Tine's Jarlsberg cheese is exported to a number of countries, and is the most widely imported cheese in the USA.

4. However, even the colour of salmon as raw material varies, and is within salmon farming a matter of scientific and practical experimentation related to content of astaxanthin in the fish feed.

5. Slinde commented in the newspapers about the hygiene conditions in the fishing industry a few years ago, accusing them of bad hygiene (*Aftenposten*, 10 February 2002; *Fiskaren*, 11 February 2004). Although claiming to argue from his own observations and research, the industry and his employer (the Norwegian Marine Research Institute) were annoyed about the statement, and defended practices that had long historical roots and that were accepted by regulation authorities.

6. See Chapter 3 for more on Tine's political role.

7. Corporate director for strategy and business development in Tine (at the time of the case study).

8. According to Berit Nordvi, around 100 persons from several organisations were counted as having made some kind of contribution in the Neptun project when the final report on the project was written in 2005.

9. Larvae of anisakid nematodes, a kind of roundworm (sometimes called whale worms), are often found in many different species of wild fish. Humans may be infected when eating raw or undercooked fish, causing anisakiasis, i.e., infections and allergic reactions. The parasite will not survive freezing down to -20 degrees Celsius for 24 hours, or warm treatment above 70 degrees Celsius. Thus, according to food regulations, fish has to be frozen before use. However, no anisakid nematodes have been found in farmed salmon, which could therefore be excepted from the regulations (Laboratory of Identification of Public Health Concern, www.dpd.cdc.gov/dpdx/HTML/Anisakiasis.htm).

10. A definition of fermentation is 'the process of deriving energy from the oxidation of organic compounds' or 'the process of energy production in a cell under anaerobic conditions (without oxygene)' (Prescott et al., 2005). Basic to fermentation in food is the work of lactic acid bacteria cultures or other micro-biological elements, such as yeast in bread and wine, and mould in blue cheese. In a salami, these active 'cultures' produce acid, increase the pH level in the product, thus conserving and curing the product.

11. Ferrières (2006) has shown how the status of food has changed through history, e.g., in relation to the urbanisation of Europe during the medieval centuries. In the rural and pre-industrial farmer societies, cured meat had higher status because of its superb qualities of storage. But with the growth of the European city cultures, with food markets offering fresh food, and with economic growth and increasing division of labour the foundation for valuing fresh meat was built. Without the need (or the space) for storing large amounts of food, fresh meat slowly took over as the preferred raw material for private households in urban areas. In other words, this was a movement of status, from 'low' to 'high', and back again.

12. The 'filter fermentor' was a 4 x 4 x 3 metre cube, with a lot of tubes and valves and tanks, a big closed box (the process management is automated), and a small control room with two seats, instruments, PCs, etc. A piece of technological complexity, enabling continuous processing (not 'batches') of lactic acid bacteria cultures for large-scale production. This is the only machine of this kind, the prototype. The bacteria get food from a set of containers, through automated control of the valves. It is a fascinating machine, but even more a fascinating story of innovation; combinations of incidents and qualified targeted work. This actant significantly supports Tine's successful growth with the Jarlsberg cheese in the US and in EU, and then, also, the fermentation of fish.

13. In reality, we could also say that the problem as such was a result of the creative practice of scientists. Hence, both the problem and its solution are socio-materially constructed. The only thing missing, then, was to construct its use(rs) – a demanding process of negotiating distributors' distribution portfolios and consumers' daily eating habits and food-making practices.

14. According to my informants, Tine launched around 60 new products during 2004, most of them small variations on and developments of existing dairy products. While the market for traditional dairy products (milk in particular) decreased throughout the 1980s and 90s, and then stabilised (in volume), Tine has been able to compensate by launching new and 'value-added' products and steadily increased their revenues and profits.

15. The Umi No Kami project and the fish oil/by-product factory Maritex in Vesterålen producing purified omega 3 to be added in 'functional foods'.

16. Based on Huse and Schøning's observations and interviews in their study of the Tine board of directors during 2004, in addition to my own conversations and interviews with participants.

17. As mentioned earlier, the effects of the various input factors were contested between the different groups involved. This version of the story is made by the'winning team', with links to Bremnes and the pre-rigor salmon. However, everyone I spoke to (on both sides) acknowledged *some* effect of the other group's input factors, even if their own factors were clearly rated higher.

18. 'Brown fat' is a small brown area between the belly loin and the back loin. When dried, remaining brown fat will show as distinct dark spots in the product.
19. An experienced dairy manager.
20. This is said to be the 'industry standard' in the fish industry; a maximum of 10,000 bacteria per gram when shipped out of the factory.
21. Norseland is Tine's own distribution company for cheese in the US. They have been incredibly successful with Jarlsberg cheese over more than 20 years, and it is now the most widely imported cheese in the US.
22. The name has been changed.
23. Later, I contacted MRC R&D in Singapore for their view, but they did not want to elaborate on the issue.
24. To show the intertwining of innovation processes, I have included quite a bit of the pre-rigor story in this case study. In a post-hoc rationalisation of the study, one could say that the successful pre-rigor innovation process is *the one* that should have been studied. But I have two objections to this, one principled and one empirical. First, from a position of methodological agnosticism, one cannot anticipate the outcome of a process a priori. Second, my point of access to this empirical study was via Tine; thus, I started out with field studies there, before expanding the field to include Bremnes as the actors and activities under investigation led me there. When I started the field study, the relation between pre-rigor salmon and fish salami was just one of several possibilities.
25. I have instead tried to describe these processes as the actors in my field study saw them; often framing things as continuous development, although now and then contributing to more discontinuous moves from one thing to another.

5   An analytic scheme of innovation processes

1. This term does not refer to James March's (1991) conceptual dichotomy of 'exploration of new possibilities' and 'exploitation of old certainties' related to organisational learning.

6   The contrary forces of innovation

1. The term 'framing' is borrowed from Goffman, and used by, e.g., Callon (1999) as a conception of how economic theory and others are used to draw boundaries around and provide the performative principles of a certain practice. In this way, the practising of markets and the theory of markets form a mutually constitutive relationship. Drawing on Bijker et al. (1987), Garud and Karnøe (2001: 10) describes frames as 'a set of beliefs, standards of evaluation, and behaviours'.
2. See, e.g., Bijker and Law (1992) for more on the notion of 'interpretative flexibility'.
3. See Hoholm and Mørk (2009) for a discussion of knowing, learning and innovating from a 'practice-based' and networked perspective.

4. With a 'flat ontology', as in actor-network theory, all action is considered local. It is the linking of locales, making networks hold together across time and space, that make up the 'macro-actors' that in other theories are stratified into different 'levels'. See also Chapter 2 on methodology.

5. Håkansson and Waluszewski (2001a; 2001b) argue that the more common concept of 'inertia' is of little help understanding the dynamic aspect of such networks, due to its fundamentally static meaning (inertia originally describes a situation where the object is not under pressure from any source, thus either standing still or moving with constant speed).

References

Akrich, M., Callon, M. and Latour, B. (2002a) 'The Keys to Success in Innovation, Part I: The Art of Interessement', *International Journal of Innovation Management*, 6(2).

Akrich, M., Callon, M. and Latour, B. (2002b) 'The Keys to Success in Innovation, Part II: The Art of Choosing Good Spokespersens', *International Journal of Innovation Management*, 6(2).

Araujo, L. (1998) 'Knowing and Learning as Networking', *Management Learning*, 29(3), Sage.

Araujo, L. (2003) 'Technological Practice, Firms, Communities and Networks', paper presented at the IMP-conference 2003, www.impgroup.org

Araujo, L. (2007) 'Markets, Market Making and Marketing', *Marketing Theory*, 7(3), Sage.

Aslesen, H. W., Mariussen, Å., Olafsen, T., Winther, U. and Ørstavik, F. (2002) 'Innovasjonssystemet i norsk havbruksnæring', The STEP Report Series, R-16, Oslo.

Awaleh, F. (2008) 'Interacting Strategically within Dyadic Business Relationships: A Case Study from the Norwegian Electronics Industry', PhD dissertation, BI Norwegian School of Management.

Azimont, F. and Araujo, L. (2007) 'Category Reviews as Market-shaping Events', *Industrial Marketing Management*, 36, Elsevier.

Bakken, T. and Hernes, T. (2006) 'Organizing is Both a Verb and a Noun: Weick meets Whitehead', *Organization Studies*, 27(11), Sage.

Barry, A. and Slater, D. (eds) (2005) *The Technological Economy*, Routledge.

Beunza, D. and Stark, D. (2004) 'How to Recognize Opportunities: Heterarchical Search in a Trading Room', in K. Knorr Cetina and A. Preda (eds), *The Sociology of Financial Markets*, Oxford University Press.

Bijker, W. (1995) *Of Bicycles, Bakelites, and Bulbs: Toward a Theory of Sociotechnical Change*, MIT Press.

Bijker, W. and Law, J. (eds) (1992) *Shaping Technology, Building Society: Studies in Sociotechnical Change*, MIT Press.

Bijker, W., Hughes, T. P. and Pinch, T. (eds) (1987) *The Social Construction of Technological Systems*, MIT Press.

Brekke, A. (2009) 'A Bumper!? An Empirical Investigation of the Relationship between the Economy and the Environment', PhD thesis, Norwegian School of Management BI.

Brown, J. S. and Duguid, P (1991) 'Organisational Learning and Communities of Practice: Towards a Unified View of Working, Learning and Innovation', *Organisation Science*, 2(1).

Busch, L. (2004) 'New! Improved? The Transformation of the Agrifood System', *Rural Sociology*, 69(3).

Busch, L. (2007) 'Performing the Economy, Performing Science: From Neoclassical to Supply Chain Models in the Agrifood Sector', *Economy and Society*, 36(3).

Calas, M. and Smircich, L. (1999) 'Past Postmodernism? Reflections and Tentitive Directions', *Academy of Management Review*, 24(4).

Callon, M. (1986) 'Some Elements of a Sociology of Translation', in J. Law (ed.), *Power, Action and Belief*, Routledge.

Callon, M. (ed.) (1998) *The Laws of the Markets*, Blackwell.

Callon, M. (1999) 'Actor-network Theory – The Market Test', in J. Law and J. Hassard (eds), *Actor Network Theory and After*, Blackwell.

Cochoy, F. (2005) 'A Brief History of "Customers", or the Gradual Standardization of Markets and Organizations', *Sociologie du Travail*, 47, Elsevier.

Cox, J. W. and Hassard, J. (2007) 'Ties to the Past in Organization Research: A Comparative Analysis of Retrospective Methods', *Organization*, 14(4).

Czarniawska, B. (1997) *Narrating the Organization: Dramas of Institutional Identity*, The University of Chicago Press.

David, P. A. (1985) 'Clio and the economics of QWERTY', *American Economic Review*, 75(2).

Dubois, A. and Araujo, L. (2004) 'Research Methods in Industrial Marketing Studies', in H. Håkansson, D. Harrison and A. Waluszewski (eds), *Rethinking Marketing: Developing a New Understanding of Markets*, Wiley.

Edquist, C. (2001) 'The Systems of Innovation Approach and Innovation Policy: An Account of the State of the Art', paper presented at the DRUID Conference, June, Aalborg.

Espeli, H., Bergh, T. and Rønning, A. (2006) *Melkens pris – perspektiver på meierisamvirkets historie*, Tun Forlag.

Ferrières, M. (2006) *Sacred Cow, Mad Cow: A History of Food Fears*, Colombia University Press.

Ford, D., Gadde, L. E., Håkansson, H. and Snehota, I. (2003) 'Managing Networks', in D. Ford, L. E. Gadde, H. Håkansson and I. Snehota (eds), *Managing Business Relationships*, Wiley.

Fox, S. (2004) 'Communities of Practice, Foucault and Actor Network Theory', *Journal of Management Studies*, 37(6).

Garsten, C. and de Montoya, M. L. (eds) (2004) *Market Matters: Exploring Cultural Processes in the Marketplace*, Palgrave Macmillan.

Garud. R. and Ahlstrom, D. (1997) 'Technology Assessment: A Socio-cognitive Perspective', *Journal of Engineering and Technology Management*, 14.

Garud, R. and Karnøe, P. (eds) (2001) *Path Dependence and Creation*, Lawrence Erlbaum Associates.

Garud, R. and Munir, K. (2008) 'From Transaction to Transformation Costs: The Case of Polaroid's SX-70 Camera', *Research Policy*, 37.

Garud, R. and Rappa, M. A. (1994) 'A Socio-cognitive Model of Technology Evolution: The Case of Cochlear Implants', *Organization Science*, 5(3).

Geertz, C. (1973) *The Interpretation of Cultures*, Basic.

Gherardi, S. and Nicolini, D. (2002) 'Learning in a Constellation of Interconnected Practices: Canon or Dissonance?', *Journal of Management Studies*, 39(4), Blackwell.

Gherardi, S., Nicolini, D. and Odella, F. (1998) 'Towards a Social Understanding of How People Learn in Organizations', *Management Learning*, 29(3).

Grint, K. and Woolgar, S. (1997) *The Machine at Work*, Polity Press.

Gupta, A. K., Tesluk, P. E., and Taylor, M. S. (2007) 'Innovation At and Across Multiple Levels of Analysis', *Organization Science*, 18(6).

Hackett, E. J., Amsterdamska, O., Lynch, M. and Wajcman, J. (2008) *The Handbook of Science and Technology Studies*, MIT Press.

Håkansson, H. and Ford, D. (2002) 'How Should Companies Interact in Business Networks?', *Journal of Business Research*, 55(2).

Håkansson, H. and Johanson, J. (eds) (2001) *Business Network Learning*, Pergamon.

Håkansson, H. and Snehota, I. (1989) 'No Business is an Island: The Network Concept of Business Strategy', *Scandinavian Journal of Management*, 5(3).

Håkansson, H. and Snehota, I. (eds) (1995) *Developing Relationships in Business Networks*, Routledge.

Håkansson, H. and Waluszewski, A. (2001a) *Managing Technological Development: IKEA, the Environment, and Technology*, Routledge.

Håkansson, H. and Waluszewski, A. (2001b) 'Co-evolution in Technological Development: The Role of Friction', paper presented on the IMP conference in Oslo, www.impgroup.org

Håkansson, H. and Waluszewski, A. (2007a) *Knowledge and Innovation in Business and Industry: The Importance of Using Others*, Routledge.

Håkansson, H. and Waluszewski, A. (2007b) 'Interaction: The Only Means to Create Use', in H. Håkansson and A. Waluszewski (eds), *Knowledge and Innovation in Business and Industry: The Importance of Using Others*, Routledge.

Hammersley, M. and Atkinson, P. (1995) *Ethnography: Principles in Practice* (2nd edn), Routledge.

Hanseth, O., Aanestad, M. and Berg, M. (2004) 'Actor-network Theory and Information Systems: What's So Special?', *Information Technology & People*, 17(2).

Hargrave, T. J., and Van de Ven, A. H. (2006) 'A collective action model of institutional innovation', *Academy of Management Review*, 31: 864–888.

Harrison, D. and Laberge, M. (2002) 'Innovation, Identities and Resistance: The Social Construction of an Innovation Network', *Journal of Management Studies*, 39(4).

Hernes, T. (2004) *The Spatial Construction of Organization*, John Benjamin Publishing.

Hernes, T. (2007) *Understanding Organization as Process: Theory for a Tangled World*, Routledge.

Hess, D. J. (1992) 'The New Ethnography and the Anthropology of Science and Technology', *Knowledge and Society: The Anthropology of Science and Technology*, 9.

Hoholm, T. and Mørk, B. E. (2009) 'Stabilizing New Practices: A Case Study of Innovation and Networked Learning', paper presented at the International Conference for Organizational Learning, Knowledge, and Capabilities, Amsterdam 27–28 April.

Holm, P. (1999) 'Fisheries Resource Management as a Heterogeneous Network', paper presented at the Eco-Knowledge Working Seminar, Nova Scotia, May.

Howard-Grenville, J. A. and Carlile, P. R. (2006) 'The Incompatibility of Knowledge Regimes: Consequences of the Material World for Cross-domain Work', *European Journal of Information Systems*, 15.

Huse, M. and Schøning, M. (2005) *Corporate governance og prosessorientert styrearbeid: Evaluering av konsernstyret i Tine BA* (Evaluation of the Corporate Board at Tine BA), Research Report No.1-2005, BI Norwegian School of Management.

Johanson, J. and Mattsson, L. G. (1987) 'Interorganizational Relations in Industrial Systems: A Network Approach Compared with the Transaction Cost Approach', *International Journal of Management and Organization*, 17(1).

Karlsen, B. (2006) 'Organisatoriske valg: Etablering og utvikling av nye arbeidsformer offshore' (Organizational Choices: Establishing and Developing New Work Practices Offshore), PhD thesis, University of Oslo.

Kjellberg, H. and Helgesson, C.-F. (2006) 'Multiple Versions of Markets: Multiplicity and Performativity in Market Practice', *Industrial Marketing Management*, 35.

Kjellberg, H. and Helgesson, C.-F. (2007a) 'The Mode of Exchange and Shaping of Markets: Distributor Influence in the Swedish Post-war Food Industry', *Industrial Marketing Management*, 36, Elsevier.

Kjellberg, H. and Helgesson, C.-F. (2007b) 'On the Nature of Markets and their Practices', *Marketing Theory*, 7(2), Sage.

Kline, S. J. and Rosenberg, N. (1986) 'An Overview of Innovation', in R. Landau and N. Rosenberg (eds), *The Positive Sum Strategy*, National Academy Press.

Knorr Cetina, K. (1981) *The Manifacture of Knowledge*, Pergamon Press.

Knorr Cetina, K. (1999) *Epistemic Cultures: How the Sciences Make Knowledge*, Harvard University Press.

Knorr Cetina, K. (2001) 'Objectual Practice', in T. Schatzki, K. Knorr Cetina and E. von Savigny (eds), *The Practice Turn in Contemporary Theory*, Routledge.

Knorr Cetina, K. and Preda, A. (eds) (2005) *The Sociology of Financial Markets*, Oxford University Press.

Knutsen, H. (ed.) (2007) *Norwegian Agriculture: Status and Trends 2007*, report from Norwegian Agricultural Economics Research Institute.

Latour, B. (1987) *Science in Action: How to Follow Scientists and Engineers Through Society*, Harvard University Press.

Latour, B. (1988) 'The Powers of Association', J. in Law (ed.), *Power, Action and Belief*, Routledge.

Latour, B. (1992) 'Where are the Missing Masses? A Sociology of a Few Mundane Artefacts', in W. Bijker and J. Law (eds), *Shaping Technology, Building Society: Studies in Sociotechnical Change*, MIT Press.

Latour, B. (1996) *Aramis, or the Love of Technology*, Harvard University Press.

Latour, B. (1999a) 'On Recalling ANT', in J. Law and J. Hassard (eds), *Actor-Network Theory and After*, Blackwell.

Latour, B. (1999b) *Pandora's Hope: Essays on the Reality of Science Studies*, Harvard University Press.

Latour, B. and Woolgar, S. (1979/1986) *Laboratory Life: The Construction of Scientific Facts*, Princeton University Life.

Law, J. (1992) 'Notes on the Theory of the Actor-Network: Ordering, Strategy and Heterogeneity', *Systems Practice* 5(4).

Law, J. (1994) *Organizing Modernity*, Blackwell.

Law, J. (1999) 'After ANT: Complexity, Naming and Topology', in J. Law and J. Hassard (eds), *Actor-Network Theory and After*, Blackwell.

Law, J. (2004) *After Method: Mess in Social Science Research*, Routledge.

Leek, S., Turnbull, P., and Naude, P. (2003) 'Interactions, Relationships and Networks: Past, Present and Future', paper presented at the IMP conference 2001, www.impgroup.org

Lundvall, B. Å., Johnson, B., Andersen, E. S. and Dalum, B. (2002) 'National Systems of Production, Innovation and Competence Building', *Research Policy*, 31, Elsevier.

Lury, C. (2004) *Brands: The Logos of the Global Economy*, Routledge.

Lynch, M. (1985) *Art and Artifact in Laboratory Science: A Study of Shop Work and Shop Talk in a Research Laboratory*, Routledge and Kegan Paul.

MacKenzie, D. (2006) *An Engine, Not a Camera: How Financial Models Shape Markets*, MIT Press.

March, J. (1991) 'Exploration and Exploitation in Organizational Learning', *Organization Science*, 2(1).

Mattsson, L. G. (2003) 'Understanding Market Dynamics: Potential Contributions to Market(ing) Studies from Actor-Network Theory', paper presented at the IMP Group Conference, Lugano.

Mattsson, L. G. and Johanson, J. (2006) 'Discovering Market Networks', *European Journal of Marketing*, 40(3/4).

McMullen, J. S. and Shepherd, D. A. (2006) 'Entrepreneurial Action and the Role of Uncertainty in the Theory of the Entrepreneur', *Academy of Management Review*, 31(1).

Medlin, C. (2002) 'Interaction: A Time Perspective', paper presented at the IMP Group Conference, www.impgroup.org.

Michael, M. (1996) 'Constructing Actor Network Theory', in M. Michael, *Constructing Identities: The Social, the Non-human and Change*, Sage.

Ministry of Fishery and Coastal Affairs and Ministry of Agriculture and Food (2004) *Den blågrønne matalliansen: Samlet innsats of ny struktur*, ('The Bluegreen Food Alliance: Joint Efforts and New Structure'), recommendation from the steering committee appointed by the Norwegian government.

Mørk, B. E. and Hoholm, T. (2008) 'Changing Practice through Boundary Organizing: A Case from Medical R&D', paper presented at the International Conference for Organizational Learning, Knowledge, and Capabilities (OLKC), Copenhagen.

Mørk, B. E., Hoholm, T. and Aanestad, M. (2006), 'Constructing, Enacting and Packaging Innovations: A Study of a Medical Technology Project', *European Journal of Innovation Management*, 9(4), Emerald.

Mørk, B. E., Hoholm, T., Aanestad, M., Edwin, B. and Ellingsen, G. (2010), 'Challenging Expertise: On Power Relations Within and Across Communities of Practice in Medical Innovation', *Management Learning* (online first, August 2010), Sage.

Mouritsen, J. and Dechow, N. (2001) 'Technologies of Managing, and the Mobilization of Paths', in R. Garud and P. Karnøe (eds), *Path Dependence and Creation*, Lawrence Erlbaum Associates.

Olsen, P. I. (2000) 'Transforming Economies', Norwegian School of Management, PhD dissertation series no. 1–2000.

Olsen, P. I. (forthcoming) 'The Relevance and Applicability of Process Metaphysics to Organizational Research', accepted for publication in *Philosophy of Management*.

Olsen, P. I. and Espelien, A. (2003) *Næringsutvikling i norsk jordbruksvaresektor: kommersialisering av teknologiske nisjer*, Research Report 12/2003, BI Norwegian School of Management.

Orlikowski, W. (2002) 'Knowing in Practice: Enacting a Collective Capability in Distributed Organizing', *Organization Science*, 13(3).

Oudshoorn, N., and Pinch, T. (2008) 'User-Technology Relationships: Some Recent Developments', *The Handbook of Science and Technology Studies*, MIT Press.

Pavitt, K. (2005) 'Innovation Process', in J. Fagerberg, D. C. Mowery and R. R. Nelson (eds), *The Oxford Handbook of Innovation*, Oxford University Press.

Pickering, A. (1995) *The Mangle of Practice: Time, Agency and Science*, University of Chicago Press.

Pinch, T. (2001) 'Why Do You Go to a Piano Store to Buy a Synthesizer?: Path Dependence and the Social Construction of Technology', in R. Garud and P. Karnøe (eds), *Path Dependence and Creation*, Lawrence Erlbaum Associates.

Pinch, T. and Oudshoorn, N. (2003) *How Users Matter: The Co-construction of Users and Technology*, MIT Press.

Pinch, T. and Oudshoorn, N. (2008) 'User-Technology Relationships: Some Recent Developments', in E. J. Hackett, O. Amsterdamska, M. Lynch and J. Wajcman (eds), *The Handbook of Science and Technology Studies*, MIT Press.

Pinch, T. and Trocco, F. (2002) *Analog Days: The Invention and Impact of the the Moog Synthesizer*, Harvard University Press.

Porter, M. (1998) 'Clusters and the New Economics of Competition', *Harvard Business Review*, November/December.

Powell, W., Koput, K. W. and Smith-Doerr, L. (1996) 'Interorganizational Collaboration and the Locus of Innovation: Networks of Learning in Biotechnology', *Administrative Science Quarterly*, 41(1).

Prescott, L. M., Harley, J. P. and Klein, D. A. (2005) *Microbiology*, McGrawHill.

Research Council of Norway (2006) *Bærekraft og innovasjon i blått og grønt: Rapport fra åtte forskningsprogrammer 2000–2005*, Research Report.

Risan, L. C. (2003) 'Hva er ei ku? Norsk Rødt Fe as teknovitenskap og naturkultur' (What is a Cow? The Norwegian Red Cow as Technoscience and Natureculture), PhD thesis, University of Oslo.

Ritter, T. and Ford, D. (2004) 'Interactions between Suppliers and Customers in Business Markets', in H. Håkansson, D. Harris and A. Waluszewski (eds), *Rethinking Marketing*, Wiley.

Rogers, E. M. (1995) *Diffusion of Innovations*, Free Press.

Scott, W. R. (1995) *Institutions and Organizations: Theory and Research*, Sage.

Spinosa, C., Flores, F. and Dreyfus, H. L. (1997) *Disclosing New Worlds: Entrepreneurship, Democratic Action and the Cultivation of Solidarity*, The MIT Press.

Star, S. L. and Griesemer, J. R. (1989) 'Institutional Ecology, "Translations" and Boundary Objects: Amateurs and Professionals in Berkeley's Museum of Vertebrate Zoology, 1907–39', *Social Studies of Science*, 19.

Stark, D. (2006) 'For a Sociology of Worth', paper presented at the European Group of Organization Studies (EGOS), Bergen.

Törmänen, A. and Möller, K. (2003) 'The Evolution of Business Nets and Capabilities: A Longitudinal Study in the ICT Sector', paper presented at the IMP Group Conference, Lugano.

Tsoukas, H. and Chia, R. (2002) 'On Organizational Becoming: Re-thinking Organizational Change', *Organization Science*, 13(5).

Van de Ven, A. and Poole, M. S. (2005) 'Alternative Approaches for Studying Organizational Change', *Organization Studies*, 26(9), Sage.

Van de Ven, A., Polley, D. E., Garud, R. and Venkataraman, S. (1999) *The Innovation Journey*, Oxford University Press.

Van Maanen, J. (1988) *Tales of the Field*, University of Chicago Press.

Vickers, D. and Fox, S. (2005) 'Powers in a Factory', in B. Czarniawska and T. Hernes (eds), *Actor Network Theory and Organizing*, Liber.

Von Hippel, E. (1988) *The Sources of Innovation*, Oxford University Press.

Von Hippel, E. (1998) 'Economics of Product Development by Users: The Impact of "Sticky" Local Information', *Management Science*, 44(5).

Von Hippel, E. (2005) *Democratizing Innovation*, MIT Press.

Walsham, G. (1995) 'Interpretive Case Studies in IS Research: Nature and Method', *European Journal of Information Systems*, 4.

Waluszewski, A. (2004) 'A Competing or Co-operating Cluster or Seven Decades of Combinatory Resources? What's Behind a Prospering Biotech Valley?', *Scandinavian Journal of Management*, 20: 125–50.

Weick, K. (1979) *The Social Psychology of Organizing*, Random House.

Weick, K. (1989) 'Theory Construction as Disciplined Imagination', *The Academy of Management Review*, 14(4).

Weick, K. (1995) *Sensemaking in Organizations*, Sage.

Woolgar, S. (1991) 'Configuring the User: The Case of Usability Trials', in J. Law (ed.) *A Sociology of Monsters*, Routledge.

Index